Revelations

by

Harold W.G. Allen

Perspective Books

Revelations

by

Harold W.G. Allen

By the same author:

Ye Shall Know the Truth
The Edge of the Universe
Higher Perspective
The Eternal Universe
The Face on Mars
Cosmic Perspective
Lunar Encounter
Asteroid
The New Cosmology
In Search of Reality

Second Edition

Perspective Books
P.O. Box 6123
Spring Hill, FL., 34611
U.S.A.

ISBN: 978-09624555-7-1

Printed in the U.S.A.

Dedication

It is a feature of our earthly existence that one may strive a lifetime in a vain quest for knowledge; while another may be so fortunate as to inherit vital information in the form of history and tradition. Scientific advances of the present generation alone have given us much precious insight that was denied to many inquisitive minds of the past. Only in a limited sense, therefore, may this work be looked upon as a personal interpretation of our rather mysterious universe. Indeed, it was made possible solely through the unselfish contributions of others, some of whom received not even the satisfaction of knowing the real worth of their discoveries.

Accordingly, it is to the memory of those who have sought in vain, to those that today and in the future are genuine in their search for Truth, that this book is hereby dedicated.

H.W.G.A.

Table of Contents

Part One
Scientific Revelations

Part Two
Religious Revelations

Part Three
Social and Economic Revelations

Note:
References are marked [1], [2], [3], etc., and are listed at the end of each chapter.

List of Illustrations

Preface

A constant threat to scientific progress is the human sin of prejudice— a trait from which no researcher can be said to be completely immune. Unfortunately, it is all too easy for a budding professional to subscribe to popular ideas, rather than risk censure from one's colleagues. But the measure of a true scientist lies in the ability to examine all possibilities— even if it means questioning the validity of "established" precepts.

At the frontiers of astronomical inquiry, where interpretation is often a matter of opinion, the temptation to jump to a conclusion is frequently very strong, as man endeavors to formulate some picture of a highly complex and mysterious universe. As a result, a number of vital concepts (such as the Big Bang model of creation) have come to be looked upon by many as actual fact instead of mere hypothesis. There is now every indication that the time is ripe for a more critical evaluation of the evidence at hand.

Prevailing religious and political beliefs will also be seen to require serious consideration and revision. For centuries, theologians have thrust a multitude of unfounded and highly illogical ideas upon a trusting and unsuspecting populace. Likewise, in their greed for wealth and power, government leaders have been guilty of distorting facts relating to basic economic principles. Their short-sighted policies have led to devastating wars which could have been avoided with sufficient knowledge.

Hopefully, with an open mind and fresh insight into such issues, the inquisitive nature of man will allow him to draw ever closer to the illusive goal of Reality: Truth.

H.W.G.A.

About the Author:

One might well say that Harold W.G. Allen's writing career started by accident. Born in Toronto, Canada (1928), he was poised to graduate from high school in close to record time when injured in a football game. Unbeknown to him, the top vertebrae in his neck had been severely wrenched out of position, interfering with the transmission of vital nerve impulses from the brain. Within a few months he was very nearly a cripple and suffering from multiple symptoms which baffled medical doctors. Forced to quit school at the age of fifteen, a proper diagnosis— and eventual cure at the hands of a chiropractor— was not made until some seven years later. During this interval, when prognosis seemed quite grim, the question of whether there was life after death became foremost in his mind.

Of an inquisitive nature, he was to begin a lifelong study into many facets of science, religion and philosophy. Interested in astronomy from an early age, he commenced a course of study in this field— only to be shocked at the incredible bias that existed in favor of a Big Bang creation cosmology. When his perfectly valid paper— "A Revised Concept of Radiation Propagation"— was unjustly rejected for publication (or even discussion) on grounds that it contradicted Einstein's relativity (a point which he was trying to make), he decided to withdraw and pursue further studies on his own. (Incredibly, his very formula— which extended the "law of addition of speeds" to radiation— would later be adopted by science in an attempt to explain how it is possible to detect redshifts in excess of light speed!)

Surprisingly, this rejection soon turned out to be a blessing in disguise. Free to tackle the problem of how a *Purposeful Universe* might be reconciled with biological evolution, without undue influence from unimaginative professors, he was to make a series of rather startling discoveries— terminating in a truly revolutionary theory of *Cosmic Reincarnation*. (Essentials of this research would later be published in 1984, by a subsidy publisher, under the title of *Higher Perspective*.)

Unable to find a mainstream publisher who could grasp the true significance of his discoveries, he eventually felt compelled to publish ongoing research under his own imprint of Perspective Books. The author's next major scientific thesis was *The Eternal Universe* (1989). This book was praised as "a startling book to treasure" by *The Observatory*, an English astronomy journal.

Some years later (1998) essentials of *The Eternal Universe,* along with his research into biblical symbolism and the Dead Sea Scrolls, were incorporated into a beautiful hardcover book: *Cosmic Perspective*. Really two books in one, in this work (published in conjunction with Sunstar Publishing) he sought to expose certain misconceptions in both science and religion. (Having given a number of public lectures on biblical symbolism, which seemed to leave some in a state of vacuum, he did not wish to undermine a reader's religious views without providing a viable alternative—hence, the two-books-in-one format.)

After publication of two science/fiction/philosophy books, *The Face on Mars* (Sunstar, 1997) and *Lunar Encounter* (Perspective Books, 2000), his next important contribution to science was to be a hardcover work entitled *The New Cosmology* (2002). This book (updated by way of an expanded softcover edition a few years later) presented startling new proof of our close proximity to a rather fateful cosmic edge of the universe.

Confronted with the spectacle of a biased scientific community that was able to suppress publication of papers critical of Big Bang cosmology, along with a reluctance of mainstream publishers to consider works contrary to the views of supposed "authorities," the author decided to establish a non-profit foundation in order to make his research available to the public. To this end, the *Cosmic Perspective Foundation* continues to mail complimentary copies of his books to selected college/university and public libraries. In a further attempt to publicize his ongoing research, the author established a Web site (www.cosmicperspective.com), as a medium through which new ideas might be freely disseminated.

Part One of his latest literary endeavor, *In Search of Reality*, conveys basic essentials of the author's earlier work, *The New Cosmology*. Although contrary to the views of many mainstream astronomers and physicists, who would continue to resist admission of error, it may be truthfully stated that no valid criticism of his revolutionary treatise has ever surfaced— in spite of repeated efforts to engage in discussion and debate. (It would seem that when no reasonable objection to new ideas can be made, silence is all that remains to preserve one's endangered credibility.)

Fortunately, some astute scientists are beginning to see merit in the "New Cosmology," and it may not be long until force of evidence and peer pressure from a growing number of unbiased researchers will succeed in reversing such adherence to past cosmological misconceptions.

Indeed, the author's research has already drawn praise from a number of reputable review sources:

"An excellent dissertation on cosmology and philosophy"—
Joseph H. Pierre, Reviewer, Amazon. com

"A powerful metaphysical study ... Exciting, fascinating, thought-provoking"— *The Midwest Book Review*

"Scientific and layman understandable"—
Mindquest Recommendations

Now, approaching his twilight years, it is the author's fervent hope that a lifetime of study and contemplation will prove helpful to others genuine in their quest for a more realistic perspective of life, and of the universe as a whole.

Part One:

Scientific Revelations

The Universe

Through distant ages man has gazed, with reverence, into the sky;
Foremost in his mind, that momentous question: Why?
How far are yonder points of light?
Shining like tiny beacons on the darkest night?
Whence came the stars, our heavenly host,
Their soft light shimmering like a phantom ghost?

Born of countless suns our Milky Way doth shine,
An enigma to ancient theories, which man must now refine.
Beyond the reaches of our Galaxy, in the enveloping depths of space,
Unnumbered starry systems participate in a truly cosmic race.
For the very universe is expanding, at a most prodigious rate;
Motivated, one would deduce, by Principle: A synonym for Fate.

H.W.G.A.

1

Evolution and Reincarnation

The Progression of Life— The Mechanism of Evolution— Universal Reincarnation

Although modern theories of stellar evolution are only able to define broad limits with regard to the age of the Earth, there is fortunately a way to obtain a date which is more accurate. In a sense, certain rocks constituting our planet's crust may be said to contain a sort of internal clock, one capable of running for billions of years and remarkably trustworthy. These rocks contain radioactive elements which decay at a specific rate. Through study of such rocks, scientists have deduced that the Earth's solid crust was likely formed some 4.5 billion years ago.

Conditions on our planet's surface were quite different in the very remote past. In particular, free oxygen is believed to have been a relatively scarce commodity, only obtaining its present abundance some billions of years later as a consequence of the eventual build-up of oxygen-secreting plants and algae. Lacking an oxygen-ozone atmospheric shield, with which to cut off much of the ultraviolet energy of the Sun, intense radiation would have bathed the surface of the Earth to a degree highly detrimental to advanced forms of life. By a remarkable coincidence, this is precisely what nature required in order to facilitate development of the earliest living structures. Even at this initial stage of evolution, one is caused to reflect whether plain chance happened to assist molecular chemistry— or whether it was really an intrinsic phase in some *Purposeful Cosmic Plan.*

The Progression of Life

Virtually all knowledge of ancient life stems from the finding of fossils, which may be defined as the remains or impression of some plant or animal that lived during a past age. A reasonable idea as to the probable age of a fossil may be obtained from a determination of the age of the rocks in which it is found. (See Fig. #1 for the divisions of geological time and a sample of life characteristic of each era.)

By the dawn of the Cambrian period, some 570 million years ago, a wide variety of marine life had become firmly established. Large numbers of trilobites, a sea-bottom inhabitant bearing a slight resemblance to a horseshoe crab, are found embedded in rocks of this particular time. Other common fossils include many types of mollusks, crustaceans and brachiopods. In fact, all major invertebrate phyla are fully represented at this stage of evolution. The advance to an internal bony skeleton is generally believed to have commenced at the beginning of the following Ordovician period, about 500 million years ago, as fossil remains of primitive vertebrates first begin to appear in rocks of this date. No trace of land life is yet in evidence, as the sea retains its exclusive franchise.

This state of affairs was to be rectified during the subsequent Silurian period, as certain forms finally commenced a transition from an aquatic environment between 450 and 400 million years ago. Elementary types of fishes also made their appearance at this time, achieving great diversification and prominence by the close of the following Devonian period, about 350 million years from the present.

While some species of early fish were able to develop external armor of solid bone, it was not until later that fish in general were able to turn their internal skeletons from cartilage into bone. It would seem that the more “progressive” amphibians, undergoing simultaneous evolution, were able to acquire bony vertebrae before the fishes themselves. In any event, the Silurian period was characterized by a number of subtle structural changes in emerging life forms— including many features which would one day provide a basis for the ultimate echelons of biological manifestation. Small fern-like plants and a variety of swampy growths were among the first vegetation to have stepped ashore, to be followed by an assortment of primitive insects, some of which resembled the centipedes and millipedes of today. The land was at last beginning to bring forth fruit.

The Devonian period witnessed the culmination of a number of revolutionary changes that had begun millions of years earlier. Certain advanced forms of marine life managed to supplant their gills with an entirely new mechanism permitting them to take oxygen from the air; they had developed lungs. (It is instructive to note that several species of lungfish, which can breathe by means of either gills or lungs, remain even today as a reminder of a past age when life was emerging from the sea. Such forms also began to convert fins into leg-like appendages, endowing their owners with the ability to crawl.) The great climax to all these elaborate modifications, which had been taking place in the sanctuary of the numerous shallow inland seas of this era, was eventually reached when the first amphibians successfully invaded the land. This new and higher life form possessed the ability to choose between a terrestrial or aquatic environment. In effect, the amphibians were now free to share the advantages and challenges of two distinct worlds— with this most recent acquisition literally opening the door to another universe.

Very noticeable changes occurred during the Carboniferous period, which began about 350 million years ago, and in the succeeding Permian period commencing roughly 70 million years later. Gigantic ferns, huge horsetails, club mosses, and dense forests were typical of this age of lowlands and tropical swamps. Large amounts of decaying matter were soon covered with water and, robbed of the oxygen of the air, eventually gave rise to deposits rich in hydrocarbons: coal and oil.

With the passage of time the amphibians were superceded by a variety of reptilian creatures with superior size, speed, and advanced egg-laying capability. Unlike their predecessors, whose eggs required constant immersion in water, they were able to produce eggs having a tough outer skin or shell which inhibited dessication. Only at this point could it be said that evolving animal phyla had truly severed the last remaining link with its aqueous origin. Appearing initially in the Carboniferous period, it was not until the following geological period that reptiles attained wide diversification and dominance. It is of note that the fossil record of this era also reveals some evidence of archaic forms of pre-mammalian life, suggesting an ancestry for mammals dating well back to the earliest reptilian inhabitants.

Spanning an interval of approximately 100 million years (from about 230 to 130 million years ago), the Triassic and Jurassic periods account for some truly bizarre fossil specimens. Often referred to as the Age of Reptiles, the size and structure of some of these creatures almost defy description. Unquestionably, the most publicized types comprise the group of so-called dinosaurs, certain species of which grew into monstrosities as great as 100 feet or more in length and weighing over 70 tons! And yet, in spite of their enormous bulk, they possessed tiny brains out of all proportion to their size.

During the subsequent Cretaceous period (about 130 to 65 million years ago) a number of very dramatic changes occurred. In the plant world such modern trees as willows, poplars, birches and oaks began to displace the ancient cycads and conifers. Swamps and lowlands underwent a series of upheavals. But the greatest surprise is to be found in the animal kingdom, for with the close of this period the seemingly invincible dinosaurs have vanished.

Although there had been several previous abrupt gaps in the fossil record, this extinction has received the most publicity. Of all proposed explanations, perhaps the most likely may be attributed to Earth's encounter with an asteroid-sized celestial body— from whence the atmosphere became so filled with pollutants that the vegetation upon which the dinosaurs depended was seriously depleted. In any event, with a conspicuous absence of the great reptiles the tiny mammals, who had been lingering in the background, were now free to multiply and achieve dominion.

The mammals at the beginning of the Tertiary period (about 65 million years ago) were typically not much larger than a cat or dog; yet despite their size they represented an immense stride over the reptilian life that had preceded them. No longer are they vulnerable to relatively small changes in temperature and humidity, for their bodies are covered with hair or fur and their blood is of constant temperature. In proportion to size, the structure of their brain is on a vastly higher level. Instead of laying eggs they bear their young alive. But what is of great significance, they nurse them and care for them until they are old enough to fend for themselves. A new era had dawned— a *teachable* form of life had come into existence, and with it had come the truly priceless advantage of learning by imitation and communication!

Fossil remains of tiny monkeys and lemur-like creatures have been found in rocks of Eocene date, as early as some 40 million years ago. Later rocks reveal increasingly larger and more varied species of primate, some of which bear a much closer resemblance to man than do the anthropoids living today. This fact clearly indicates that mankind did not descend from an existing species, and strongly points to a common ancestry somewhere back near the dawn of the Tertiary period, some tens of millions of years ago.

A procession of biped finds, dating back millions of years, paint a rather informative picture of a steady progression of skulls housing increasingly larger brains. Once the age of a fossil falls below about 50,000 years, it then becomes eligible for dating by the carbon-14 isotope technique. Utilizing this superior method, some rather specific dates have been established for a wide range of specimens. Intermediate forms from late Neanderthal to modern sapiens, his eventual successor, have been found in a number of locations and tend to reveal a continuum with regard to both physical and cultural evolution.

With an effective means of dating many otherwise uncertain finds, a deluge of information has become available concerning that interval immediately prior to the dawn of recorded history. From a probable origin in Africa, early man left his mark— in the form of paintings— on the walls of European caves some 25,000 (or more) years ago. Sites of human occupation, of similar antiquity, have been discovered in the New World, including such locations as the Canadian Yukon, Mexico and Peru. In truth, the ancestry of man is revealed to be quite extensive— utilizing virtually the entire globe as a nursery, *and with his origin rooted at the very dawn of life!*

Unfortunately, the evolution of mankind is directly opposed to a popular religious viewpoint. By tradition, most Christian clergy have vigorously rejected evolutionary theory, steadfastly upholding a blind faith in a literal interpretation of the Old Testament account of creation. Over the past century or so this official attitude has experienced little change, even if the more enlightened members of their flocks have long since entertained grave doubts as to the wisdom of ecclesiastical authorities. Some theologians, unable to completely ignore overwhelming scientific evidence, have sought to mitigate the situation by means of a compromise. In effect, they would begrudgingly accept the reality of evolution up to a point— namely, to the extent that man himself is not included in the overall scheme. This highly arbitrary exclusion is necessary, of course, in order that they might retain what must be described as a bizarre interpretation of the Adam and Eve narrative, along with an equally bizarre "Redemption" doctrine requiring a mystical fall of man.

In addition to conclusive scientific evidence, confirming man's evolutionary heritage, there is also strong support from such sources as philosophy and common sense. From a philosophical point of view, the fossil record is one of ever-increasing complexity and intelligence. This is precisely the pattern needed to inspire contemplation of the dynamic principle of evolution— leading to the conviction that we must reside in a universe ruled by *Purpose*. Plain reasoning is capable of proving the evolution of man by the simple expedient of eliminating the sole alternative of "Special Creation." Other than by admitting the evolutionary premise of changing species, the absolutely staggering profusion of fossil entities can only be explained by postulating an *almost endless number of separate creations, extending over vast periods of time!* But if God be the Creator, why all the trials and tribulations and "experiments" before finally deciding to create the particular forms of life existing today? In terms of "Special Creation" the whole affair is obviously preposterous. Evolution— *including that of man*— is capable of providing not only a far simpler explanation, but it is the only one which makes any sense at all!

The Mechanism of Evolution

In contemplating the incredible complexity of biological life, one is immediately pressed for an explanation. How could all of these forms evolve from a microscopic origin? As amazing as this profusion of life may be, it tends to pale by comparison to what occurs daily, as tiny cells quickly grow into structures containing many trillions of cells. All of this not only takes place before our very eyes, but it does so without requiring one act of "Special Creation." Since it is obvious that these everyday "miracles" could not have occurred as a result of chance, we must admit the influence of "anti-chance," or God— an assumption which also explains the overall success of evolution in spite of numerous monstrosities and blind alleys.

While a cell contains a number of other bodies, entrusted with a variety of functions, the nucleus itself is the center of organization and government. Within this region are elongated strands of a nucleic acid molecule known as DNA. Assuming the shape of specific threadlike spiral bodies called *chromosomes*, at certain stages of cell division, these structures hold the heredity information essential for reproduction and for the overall functioning of the organism. In turn, chromosomes are comprised of numerous tiny segments of intelligence commonly referred to as *genes*— each of which may be thought of as a miniature memory storage unit or "mind." Acting in conjunction with each other, these genes are able to produce a somewhat similar molecular substance named RNA, which is capable of moving out of the nucleus and acting as a type of "template" for construction of the vital amino acids upon which life is based. On the whole, it may be said that the double helix pattern of DNA molecules— with their strings of genes— is highly analogous to a linkage of vast swarms of neurons (brain cells) that constitute our own brains. Indeed, the only real difference is likely one of *magnitude*— with these minute genes interacting together to manifest an overall living entity just as surely as our cerebral structure (of linked neurons) is able to give rise to a human mind.

Thus it will be deduced that even the most elementary cell is a life form of incredible complexity, and one which could not have arisen from mere material particles were its evolution not guided by some intelligent force— *an intangible influence which must surely be ascribed to the presence and action of spirit!* Moreover, the dynamic nature of genes may be defined in terms of aggregations of atoms manifesting (or housing) a degree of intelligence, or spirit. And yet, the premise of specific concentrations of matter manifesting spirit should not really come as a surprise; for after all, where there is life there must surely be spirit.

The means by which all living organisms have evolved into their present forms must be ascribed to the outcome of countless small mutations. Insofar as evolution is concerned, a mutation can be defined as failure of the genes, inhabiting a germ cell, to reproduce in an exact manner. No form of life is free from mutations, which appear to occur at random and are almost always harmful or quite useless variations. In general, all changes of an advantageous nature are far too gradual to be noted over short periods of time.

When one surveys the path of evolution, with its seemingly miraculous transition from the realm of tiny atomic particles to incredibly complex structures, can it be seriously doubted that *guidance* was involved? Surely, the complexity of our very bodies is sufficient to convince us that we are not the product of chance. This is a conclusion which forces us to acknowledge that some "intelligent" factor must reside within the genes and, moreover, that it is the actions and influence of this factor (spirit) which is largely responsible for the success of evolution— essentially, by means of constructive mutations. What might appear to be a matter of chance must, in reality, be ascribed to the actions of some hidden intelligence at work— *ultimately, to the influence of that Cosmic Principle we choose to call God!*

It is rather strange that a materialistic approach has dominated so many attempts to explain the mystery of evolution. What might best be referred to as the "mechanical" basis of traditional theory may be summed up under such terms as natural selection, adaptation , and so-called "chance" mutations. Upon this flimsy platform purely random or accidental mutations are held to provide the raw material for the omnipotent and deciding influence of environment, which in turn is supposed to determine whether organisms possessing these new features will be allowed to live and to multiply. This process of weeding out undesirable and unfitted forms of life is known as natural selection or survival of the fittest.

Now this is all very fine, and we must certainly admit that an adverse environment will tend to eliminate any species that is malformed and unsuitably adapted, but when we attempt to apply such arguments to the problem of evolution itself we find that, in actuality, *we have a framework capable of explaining essentially nothing!* Not only must it be conceded that advantageous mutations rarely (if ever) arise through pure chance, but the very course of evolution is plagued with an almost endless number of instances which are quite contrary to the dictates of natural selection. For example, it must be apparent that practically every phase, during the transition of life toward greater complexity, rudiments of useless or even detrimental physical forms must have arisen. One might say that some mysterious providence *foresaw*, as it were, the usefulness or *later need* of many diverse and highly complicated structural modifications! Obviously, such traditional "mechanical" theories are misleading and wholly inadequate, since they fail utterly to explain the course of evolution.

Yet another problem facing supporters of the materialistic point of view may be traced to the manifestation of *mind*. Where does this "spark of intelligence"— representing the difference between life and death— continue to come from? For some unknown reason very few serious attempts have been made to incorporate the presence of *spirit* into a theory embracing evolution. Why this has been so constitutes a great mystery; *for does not all life— including our own soul— consist of spirit?* Furthermore, if we are to believe in the existence of a future life, and since there is overwhelming evidence as to the truth of evolution, we have no alternative but to recognize the presence and interaction of spirit with *all levels of creation, from man down to the most elementary gene-complexes, and even to the ultimate depths of matter itself!*

A truly fundamental problem, which has long intrigued both scientist and philosopher, is that of explaining the transition from inanimate molecular configurations to patterns characterized by life or spirit. When does a group of atoms cease to be inorganic matter and become a lowly form of life? Was the emergence of the first life on our planet simply a fluke of nature, involving a chance "fortuitous concourse of atoms?" Or, on the other hand, was it really the inevitable result of dynamic properties inherent in the very constitution of material particles?

Laboratory experiments have left no doubt as to the ability of atoms to engage in a spontaneous generation of organic molecules. Placing various common elements in a flask containing water, and autoclaving the entire apparatus so as to ensure sterility, scientists have performed tests utilizing such energy sources as heat, ultraviolet and X-ray radiation, electric spark, etc. The results turned out to be rather impressive, as a host of organic compounds were readily synthesized. In short, a wealth of complex organic chemical compounds is bound to be the rightful heritage of all newly formed Earth-like planets. Accordingly, the widespread evolution of biological life in the universe must be a foregone conclusion.

Conceding the inherent potential of matter to initiate life, the question remains as to how particular arrangements of atoms could give rise to intelligence. Not only do we have the lowly level of microscopic gene-spirits and cellular entities to consider, but there is also the overall consciousness of the entire animal phyla— of which the human soul is just a solitary example— demanding clarification. Obviously, the whole issue hinges upon a structural positioning of atomic particles so as to form specific patterns conducive to the manifestation of spirit. Bearing in mind the non-spatial nature of spirit, it may well prove useful to visualize this association with matter in terms of a "capturing" or "housing" hypothesis. That is to say, when certain intricate "fields of influence" have come into being, as a result of the formation of appropriate molecular configurations (each atom alone displays a variety of mysterious electrical, magnetic and gravitational forces, etc.), these complex fields could be thought of as attracting and capturing a level of spirit or life. Thus the intelligence of gene-spirits would be seen to arise from specific atomic/molecular patterns; while the overall personality of a unicellular organism may be interpreted as being due to the larger scale patterning of gene-type configurations. Upon a still higher plane, featuring more advanced animal life forms, the interactions of specific patterns of brain cells are able to facilitate the capturing of spirit of steadily increasing value.

In spite of a profusion of biological forms, over the lengthy course of evolution, there is clear evidence of *guidance*— ever urging an upward development of life. This is evident at every major stage of diversification. In addition to explaining the construction of numerous exceedingly complex internal organs, which are totally inexplicable on grounds of pure chance, this premise is borne out by a remarkable similarity in the external appearance of evolved life forms. For example, one has only to look at the broad spectrum of the world's mammal population. Is it just a strange coincidence that virtually all such creatures possess one head, one mouth and nose, two eyes and ears, four appendages in the form of pairs of arms and/or legs, five fingers and five toes (or claws), etc.? Much rather, this parallelism can be deduced as a striving toward some *ultimate purpose— namely, Homo sapiens, and beyond.*

The highly diversified and somewhat haphazard path described by evolution can only be resolved upon the basis of *free will.* For it is solely by ascribing freedom of will to spirit that we are able to account logically for the numerous trials and errors of evolution. Failure to recognize free will as an intrinsic attribute of spirit is really tantamount to accusing God of outright incompetence in the creation of life! Imperfection exists in the universe simply because the many levels of spirit are less than Perfect— *not as a result of any direct action on the part of a Creator!*

Adopting this line of reasoning, the outcome of evolution may be acknowledged to hinge upon *probability.* That is to say, a specific proportion of spirits can be expected to utilize their potentialities and freedom of will in such a manner as to follow specified courses. Just as small-scale uncertainty of individual atomic particles gives way to large-scale stability when really great numbers are considered, so nature has contrived to ensure overwhelming probability that an almost infinitesimal fraction of spirit would pursue paths that would terminate in man; while the vast majority could be relied upon to diverge and lead to the creation of an extensive chain of lesser species. Subsequently, our planet's biological life pattern *may be likened unto the structure of an enormous pyramid* -- extending from myriad swarms of gene-type spirits to modern man, with his greatly reduced numbers.

The hypothesis of *Divine Guidance*, as the motivation behind the success of evolution, admittedly places the entire problem of mutations and adaptation into the realm of philosophical deduction. But this in no way negates the validity of such a concept. In fact, history is permeated with instances in which philosophy and logic have preceded a scientific explanation. Contemporary science, with its roots still embedded in archaic "materialism," is just in its infancy when it comes to comprehending *Cause or Principle*. For this reason it is sufficient to theorize the existence of a subconscious "feedback" mechanism between the gene-complexes of germ cells and the overall controlling spiritual entity. While the traditional view of natural selection may be expected to play a role in the adaptation of any species to its environment, it is also clear that beneficial mutations are— in some vital sense— most certainly *directed!* Comprised as our bodies are of several distinct levels of life or spiritual awareness, it is not surprising that one manifestation should have no direct conscious association with another. Indeed, it is as though quite different worlds are involved, with each being able to affect the other indirectly by means of diverting guidance from above— *the ultimate source of which may be ascribed to God!*

Equivalent in many respects to that of conscience in man, the use of free will by the gene-spirits of germ cells must be held largely responsible for evolutionary mutations— producing both monstrosities and the remarkably advanced form epitomized by modern man. This same principle also applies to the ability of any species to undergo environmental adaptation. What very likely transpires is that need for structural change is capable of impressing itself upon the gene-complexes of germ cells. Exerting a "pressure" for a certain mutation (if at all possible), the Darwinian concept of natural selection is then permitted to act upon favorable variations. (Examples of this adaptive capability in man range from the increased resistance of some races to specific diseases, to differences of skin pigmentation in others commensurate with their exposure to sunlight, etc.)

Notwithstanding our present inability to grasp many details of biological evolution, it is both desirable and expedient to formulate one basic conclusion: *We must presume that the fundamental nature of the cosmos os such that all life contains spirit, and that it is able to rise to higher levels!* Only by making this profound assumption can we reconcile the idea of a future life with the reality of evolution. To think that man alone is the only spirit privileged to evolve to a higher status is absurd, since it is hopelessly in conflict with our knowledge that even man himself originated from a lower form of life— from a form far lower than the simplest cell, in fact!

Universal Reincarnation

A useful definition of the term "spirit" is perhaps in order at this point. Stripped of the associated attributes of intelligence and memory, it may well be constructive to define the broad spectrum of spirit as *specific and varying segments of Perfection*, or whatever

God consists of. In essence, a spiritual entity is basically a particular *degree of unselfishness*— nothing more! It is surely this feeling of *Universal Love*— expressed as a "level of reality or righteousness"— which determines the actual status that a spirit is given, in the sense that the higher this level the more advanced its status with regard to the evolutionary structure of the universe. Thus it will be seen that spirit cannot be divided into classes or categories. On the contrary, during the long course of evolution our planet must have witnessed a steady and unbroken procession of spirit of gradually increasing value, or degree of Perfection. *Principle*, in which spirit forever seeks to equate intrinsic worth with physical status, is acknowledged to be the one dominant cosmic feature— completely transcending mundane phenomena such as space and time should this be necessary to preserve Harmony and Justice. Once again, it will be emphasized that— due to evolution— *no one spirit is able to progress to a higher level of manifestation to the exclusion of all others!*

Such assumptions raise rather pertinent questions which have been asked by philosophers over the ages. In brief, why is it that we find ourselves born into an imperfect world? Why, in fact, should any form of life find itself situated where it does? If the universe is the handiwork of an Omnipotent and Perfect Creator, why is anything at all imperfect? What, in the last analysis, is the deeper significance of ascribing free will to spirit? Invariably, the issue must hinge upon *justification of status*— in which spirit is seen to be not so much created as it is *constructed* in a lengthy step-by-step process from some infinitesimal origin. In order to acquire a higher nature, all spiritual entities must *earn or deserve* this most cherished reward. Interspersed upon myriad occasions by death, the evolution of a spirit (or soul) may only proceed on the basis of what might well be described as the principle of *Universal Cosmic Reincarnation.*

Many advantages are inherent in the concept of reincarnation. The old biblical notion that we have but one all-determining life in which to prove ourselves worthy of passing to "heaven" is seen to be quite illogical. For example, what happens to the soul of one who dies in infancy? Should we be so naive as to think that it passes automatically to a higher existence we are faced with the thought that those who live longer are unlucky, inasmuch as they have risked not making it, so to speak! Furthermore, at what specific instant of time does a child cease being a child and become responsible for its actions? Clearly, no time limit can be established. A similar parallel exists with regard to adults. Who is to say that time will not lead to profound changes in one's spiritual outlook? And yet, we know that death shows respect for neither time nor person. Are we to relegate Divine Justice to the whims of blind chance, by failing to give all individuals an equal length of time and equal opportunity to evolve spiritually?

Upon the premise of reincarnation such objections will be seen to have a simple and common solution. By acknowledging the immortality of soul, or spirit, the time limit impasse is at once removed. The mystery surrounding premature death (or any death, for that matter) no longer exists, since an appropriate rebirth will prevent the injustice which would otherwise ensue. To an astute thinker, belief in reincarnation offers the only alternative to the spectacle of a chaotic universe. But what of the laws regulating this most dynamic principle? Can this concept be reconciled with evidence revealed by science?

Upon a philosophical basis, there is a very definite connection between evolution and reincarnation. If man really did evolve, from at least a microscopic origin, then where are we to draw the line as to the specific stage at which he suddenly became blessed with the potential for reincarnation? Most certainly, precisely the same problem arises as must accompany any notion of one all-determining existence. A little reasoning is sufficient to show that the two phenomena are indeed related, *and that the answer to the one difficulty is also the solution to the other!*

Rather than strive to impose a time limit where none may be imposed, or draw a line where none can be drawn, we are led to only one sensible conclusion. Without a doubt, it simply boils down to a case of *all or nothing* with regard to the issue of a life after death. Either all forms of life are eligible for a future rebirth, or else it is a false hope that does not exist at any level— including that of man himself.

Should we embrace the logical concept of a Just Creator and a Purposeful universe— in which all levels of spirit are reborn— the question of status becomes of paramount importance. For in order to justify its actual position upon the evolutionary scale of life, a spirit must *deserve* the level which it is given. Accordingly, it can only deserve a particular status if it has had a *previous existence!* This leads, inescapably, to the conviction that all spirit must be traced to an extremely low and common origin— from whence the reincarnation process achieved wide diversification from the elementary gene-spirit entities of early life forms.

An extension of this assumption to the depths of matter is not without similar advantages in resolving the paradox of a Perfect Creator and an imperfect physical universe. As a matter of fact, considering that there is often great difficulty in distinguishing between animate and inanimate forms toward the lower extremities of life, any distinction is frequently more arbitrary than not. It is, in reality, almost tantamount to attempting to draw another line where no line may be drawn. It may well be prudent, therefore, to consider even lowly matter to be of the essential nature of spirit— with the only genuine difference between life and seemingly inert matter being *one of degree or value*.

Upon reflection, the concept of reincarnation appears to be in full agreement with a spiritualistic interpretation of matter, in spite of numerous instances of atoms and material particles being torn apart and fused together by circumstances over which they have no control. Invariably, the formation of new physical bodies must induce appropriate changes in associated spirit entities, in order that status (or degree of evolutionary progress) may again be justified. The process, in this case, is much the same as the phenomenon of birth and death in our own world. About the only difference which might be cited is that we are actually able to detect the birth of new entities in matter, whereas with the death of biological life we can see no direct proof of continued existence. It is surely a stroke of irony when the indestructibility of lowly matter may be adduced in support of the principle of spiritual immortality!

Proceeding with the fundamental assumption that spirit is capable of rising to immense heights, from the depths of obscurity, we are at once confronted with an intriguing observation. It so happens that the further one goes down the scale of life the greater the numbers encountered of the lower spiritual entities. The difference in the abundance of the many species of biological life alone is positively staggering— to say nothing of the vast multitudes of atoms comprising the simplest cell. *How could they all hope to evolve, individually, to higher levels?* For if these countless swarms of lesser spirits are blessed with the potential to eventually progress to the level of man (at least)—and they must be, since man himself is the product of a less than microscopic beginning— then does it not appear that only a fantastically small proportion could ever reach this advanced status?

There can, of course, be only one sensible conclusion: *We must deduce that many spirits of a low order may— in conformity to some cosmic law— be combined to produce fewer entities of a higher overall nature or status!* Solely by recognizing the principle of ***fusion***, as a vital and dynamic feature of the universe, may we escape an otherwise absurd situation.

The mechanism of this inferred fusion process urgently invites detailed clarification. Meanwhile, there seems no compelling reason why we could not theorize a useful picture in harmony with logic and observational evidence. (Inexplicably, the significance of this fusion principle has been overlooked by scientists and philosophers alike, with the result that a most important clue was missed— one which could have facilitated a much earlier solution to certain of nature's more treasured secrets.)

Let us begin by designating the most infinitesimal segment of spirit (or quantum of mass-energy) as an **"A."** The least degree of fusion that it may enter into is union with another entity of like value. Should the two sufficiently deserve the resulting status they will surrender original identities in order to give birth to a new and higher overall entity (to which we might assign the symbol **"B"),** of twice previous individual worth. Similarly, the fusion of two **"B"**s could be viewed as leading to the formation of one **"C,"** etc. The spiritual cosmos, upon the basis of such a doubling concept, might be pictured as a series of "steps" extending to higher levels of existence, *with each successive "step" containing an equal value of* ***"A"*** *but only half as many entities.* (So effective is this doubling process that little more than 33 "steps" is equivalent to the huge sum of 10 billion.) This is not to imply that the incidence of fusion is restricted to twofold jumps in status. On the contrary, there is every reason to presume a wide variety of intermediate complexes. For instance, the union of an **"A:** with a **"B"** must lead to the creation of an entity with a value of three **"A"**s. The fusion of a **"C"** with a **"B"** would produce a structure valued at six **"A"**s, etc.

Very frequently, in the oftentimes violent world of physical events, circumstances will produce an entity of a status that is undeserved by its recently fused components. In such an instance it must bring about the "death"— and subsequent rebirth— of the separate entities on the grounds that they are unworthy of the new status so created. Spirit of the appropriate level, itself the product of a previous disruption, would immediately inhabit the physical structure of higher value. (Time and space simply do not have any meaning to spirit lacking physical manifestation.)

By way of analogy our mysterious universe might be likened unto a vast pyramid, with numerous "steps" leading to its upper reaches. The base of the structure is wide, representing a truly enormous number of entities of low worth, tapering steadily with height toward a point signifying fewer and fewer spirits of higher and higher value. At the top of this *Great Cosmic Pyramid* we have a peak or pinnacle which might be held to depict a state of *Maximum Fusion, or Oneness with God.*

Regardless of the momentous and startling conclusion which must arise, no valid excuse can be advanced to predict an end to such an evolutionary scheme only part way up the Great Cosmic Pyramid. This being so, we are left with no alternative but to recognize the inherent right of all creation to rise to the height of Perfection, or

Total Fusion. For it must surely follow that *all spirit— man included— is destined to terminate in none other than God Himself.* Fusion into the One Body of God is invariably our glorious Destiny! Therein, may well be found both the Purpose and the Motivation behind creation.

2

Our Physical Universe

The Enigma of Matter— Our Astronomical Universe— The Expanding Universe— Quasars

Knowledge of our physical universe is being pursued by science upon two basic fronts— namely, microcosm and macrocosm. At the very frontiers of microcosm, physicists have succeeded in tapping— for better or for worse— the enormous energy of atoms. On a purely beneficial note, they have also managed to unravel the mystery of the stars, providing valuable insight into the atomic processes transpiring deep within their interiors. Astronomers and astrophysicists, concerned with the vast macrocosm universe, have thus been greatly assisted by research in a branch of science endeavoring to probe in precisely the opposite direction.

Such an instance is typical of the interrelationship which exists among many seemingly diverse fields of inquiry, and affords a graphic illustration of the folly of those who would attempt to exclude philosophical deduction from the realm of science. As long as this approach is capable of explaining phenomena, and is conducive to testing of theory with observation, it provides just as sound a basis as the circumstantial evidence commonly accepted by science.

However, before proceeding with this principle, which would seek to link *Motivation or Purpose* with creation, it is necessary to examine certain aspects of our physical universe. Only through an understanding of the laws regulating so-called material structures— ranging from infinitesimal atomic entities to giant galaxies of stars— is it possible to construct a workable hypothesis that would recognize spiritual motivation as being associated with lowly matter.

The Enigma of Matter

Throughout the ages, matter has commonly been looked upon as an inert and absolutely dead substance, and certainly as having nothing to do with the spiritual aspect of the cosmos. While, for practical purposes, this contrast may have some justification, scientific advances are tending more and more to raise concepts of matter to something above previous rock-bottom estimates— to the status of an enigma, if nothing else. For as strange as it may at first seem, the deeper science probes into the secrets of matter the less "materialistic" it appears to be.

Take, for example, a tiny molecule of oxygen in the atmosphere we breathe. Who would dream that this minuscule object, which measures only about a hundred-millionth of an inch, is constantly darting around and rebounding off other molecules with a velocity of something like one-third of a mile per second? Although many other atoms are not nearly so lively in their habits, they nevertheless are all in a constant state of agitation, ranging from high velocities for gaseous substances down to a more or less rapid vibrating motion for the most rigid solids. Who can honestly consider matter to be inert and "dead?"

From a dimensional point of view, the structure of an atom may be said to consist primarily of empty space! That is to say, its center or nucleus (manifesting positive charges of an influential force known as electricity) is about 10,000 times smaller than the diameter of the atom itself, which is extended by a specific number (depending upon the element in question) of lighter concentrations of mass termed electrons. Bearing negative charges of electricity these tiny electrons revolve, so to speak, around the nucleus. Roughly analogous to incredibly minute solar systems, a major difference lies in the realization that electron orbits are constantly shifting so as to envelop the central nucleus in a series of concentric shells. Further evidence has resulted in the hypothesis of electron spin, in which these tiny particles are seen to resemble rapidly spinning tops— in addition to behaving like miniature magnets by virtue of a negative electric charge.

The chemical properties of all the various elements can be explained on the basis of the number of electrons in the outermost shell. For instance, an atom that requires one more electron to complete its outer shell will readily form a strong bond with another having a solitary electron in its most distant shell— thus neutralizing each other by reason of their ability to share an outer electron. In turn, the number of electrons (with negative charge) in a given atom will depend upon the quantity of protons (with positive charge) in the nucleus, since it is a cardinal rule of nature that unlike electrical charges attract while like charges repel. The mystery of how multiple protons could reside together in close proximity was solved with the discovery of the neutron, which has a mass slightly greater than the proton. Neutralizing positive charges, it is able to restore balance within a compound nucleus.

Instead of our being able to resolve matter into a few inert and unchangeable substances, we now find that the universe is built out of what might be described as divisible and interchangeable concentrations of energy or forces— of *"fields of influence"* which act at a distance. It has been said that electricity is a basic commodity out of which our universe is made. But just what is electricity? Moreover, what is the equally enigmatic force of gravitation? Perhaps the conspicuous failure of science to resolve such mysteries is an indication that an entirely new approach is in order— presumably, one in which *Principle* is seen to play a truly dominant role.

To summarize, it may be stated that the two fundamental particles of matter, which alone possess the ability to exist in a free and stable state, are the proton and electron— or their oppositely charged counterparts, the antiproton and positron. Extremely rare in our region of the universe, these counterparts have only a very transient existence, being promptly annihilated in collisions with common matter. For some very good reason, which must be ascribed to circumstances peculiar to creation, virtually no free positrons or antiprotons have been permitted to inhabit our region of space. This is just as well, of course, since their presence would lead to a chaotic annihilation of material particles and a universe consisting of little else than pure energy. Invariably, this biased production and distribution of fundamental particles must be given due consideration by any theory attempting to account for the origin of our physical universe.

Our Astronomical Universe

As we all know, the Sun occupies the central position in our Solar System. Essentially a huge ball of intensely hot gas, it has a diameter of about 864,000 miles and weighs roughly one-third of a million times more than the Earth. Although some 93 million miles distant from us, it is not its particulary high luminosity but rather its comparative nearness which is responsible for appearing so much more brilliant than the stars. Indeed, as great a contrast as there may appear to be, our Sun is simply an average star. It just happens to be the closest.

Since one of the original discoveries about our Solar System was the knowledge that all of the planets revolve in the same direction as the Sun rotates, and on virtually the same plane as the Sun's equator, it was inevitable that this remarkable coincidence should inspire an idea as to the origin of our family of planets. Accordingly, it was not long before theory was advanced to the effect that the Sun and planets were formed, at about the same time, from the condensation of a cloud of interstellar gas and dust. In conformity to a firmly established principle of physics, whereby contraction of a rotating body must lead to an increase in its rate of spin, it was assumed that when our infant sun had shrunk to a critical degree, strong rotary forces must have caused periodic ejection of material— from whence the planets and their moons later condensed.

Although rather confused for a while by a supposed problem that involved conservation of angular momentum, it is now known to be essentially correct. Consequently, the formation of a family of planets is seen to be the birthright of most newly formed stars. While it is quite evident that many of these potential abodes for biological life will be wasted, for one reason or another, it is impossible to escape the conviction that at least a reasonable proportion of them will produce planetary systems comparable to Earth. The total number of favorable abodes for life must be truly astronomical, given the vast multitudes of stars and galaxies in the universe. As to the question of whether life similar to our own will develop on these other worlds, we have only to ponder the notions of certain misguided ancestors who once thought in terms of a flat Earth situated at the center of the universe. Clearly, this new line of reasoning must dominate future philosophical inquiry into the nature of the cosmos.

Moving away from the confines of our Solar System, let us turn to the highly fascinating subject of the stars themselves. The method used to determine the distances of the nearest stars is simple in principle. As a result of the Earth's revolution around the Sun, nearby stars will appear to describe a slight backward and forward motion with respect to more distant background stars. Utilizing this annual displacement (or "parallax"), it can be shown that even the closest star is quite remote— some 25 million million miles away. In order to more conveniently express such distances, the term "light-year" has been adopted. A light-year depicts the distance covered in one year by a ray of light traveling at the speed of some 186,000 miles per second. In this manner, the nearest star has been found to be 4.3 light-years away; while Sirius, the brightest star in the sky, is at a distance of about 8.6 light-years.

Observation reveals a great variation in both size and density of the stars. While most do not differ too greatly in diameter and mass from our Sun and belong to what is known as the "main-sequence" (e.g., stars that have undergone little evolutionary change), there are some so large that if one of them were to replace the Sun, the orbits of the inner planets would actually lie within the body of the star itself. Such a typical "red giant" star may be more than a thousand times as luminous as the Sun. At the other extreme are stars so small that while they possess the mass of stars they resemble our planet in size. Objects of this category are referred to as "white dwarfs." As if this is not enough, even more compacted structures have been found to exist in the form of "neutron stars," typically about ten miles in diameter and several times the mass of the Sun!

By comparing the luminosities of the stars with their masses, it was discovered that the heavier stars were consuming their fuel at a rate far higher than average. While most stars are capable of burning for billions of years, the more massive ones will exhaust their supply of hydrogen fuel in mere millions of years, and so meet their fate in just a fraction of the lifetime of the average star. To grasp the sequence of events which link bloated red giant stars with the crushed state of the white dwarfs, it is necessary to understand what determines the size of a newly formed star. The diameter of a star is the outcome of two opposing forces. On the one hand, we have the force of gravitation which would attempt to draw all matter into a common center of infinite density. Opposed to this unifying influence is the explosive or repulsive force of heat— in essence, motion which acts to drive the material of a star apart. Condensing out of a hydrogen-rich gas cloud, a newborn star will cease contracting when sufficient heat is generated, deep within its central core, to support the weight of overlying layers. At this point it will stop shrinking and become a normal main-sequence star just like our Sun, with its temperature being a distinct function of mass— in the sense that massive stars are the hottest and will burn their fuel at a much faster rate.

With time, the core of a star will continue to collapse and to generate higher temperatures as nuclear reactions increase. This induces the outer atmosphere of a star to expand—causing it to move off the main-sequence en route to the realm of the inflated red giants. While it has taken our Sun roughly five billion years to evolve to its present status, which is only slightly removed from its original state, one more such interval will produce very profound changes.

What happens after a critical proportion of a star's hydrogen is consumed? Upon reaching the stage where the generation of nuclear energy is insufficient to support a star's outer regions, expansion is reversed and a degree of gravitational collapse is initiated. As a result of this contraction the star will temporarily become heated and undergo expansion (perhaps ejecting a small portion of its atmosphere) until subsequent cooling permits another contraction. In effect, it may now be transformed into a pulsating star, of which the Cepheids are an example of one particular class or stage of evolution. (Since all pulsating stars are of high luminosity, they became of great value to astronomers once it was noted that their pulsation rate was closely linked to intrinsic luminosity— a feature which has allowed great distances to be measured.) Contracting far beyond its initial state on the main-sequence, such a star may end up as a small high-density white dwarf star.

Not all stars are privileged to follow this evolutionary loop without interruption. Extreme violence lies in store for the more massive stars. Long before a really heavy star is able to consume all of its hydrogen, the helium core will reach temperatures that are conducive to the formation of more complex elements. Helium is converted into such elements as carbon, oxygen, neon, etc. In due course, heavier and heavier elements are built up in layers within the star, with the ultimate iron group of heavy metals at the center arising from the burning of the silicon group which had preceded it. Once a critical mass and temperature is reached, electrons are squeezed inside the heavy nuclei where they unite with oppositely charged protons to form neutrons. Occupying less space than that of their combined separate identities, the collapse of the core is accelerated— resulting in even greater numbers of neutrons being formed and an even faster rate of contraction. This eventually leads to a runaway collapse and a catastrophic implosion of the core. The enormous energy suddenly released is sufficient to detonate all of the remaining unburned elements in the doomed star— producing a monstrous explosion known as a "supernova," in which radiation is released equal to the light of hundreds of millions of suns! Having thus rid itself of much of its mass, the stellar remnant is allowed to pursue evolution to its predestined status of a neutron star, or beyond.

As dramatic as a supernova may be, it is not a phenomenon without some cosmic purpose. Were it not for these stupendous explosions, biological life would have been quite impossible. This is because all the heavier elements are formed in the interiors of stars, especially massive stars destined to become a supernova. It is only through explosions of extreme violence that these vital ingredients manage to intersperse with clouds of otherwise almost pure hydrogen gas, so that later generations of stars are able to condense out of just the required medium to give birth to planets with the proper chemical composition. Of the matter comprising our very bodies, much of it was once part of a supernova!

The eventual fate of a degenerate star hinges upon the crucial issue of mass, which allows but three choices: white dwarf star, neutron star or black hole. Stability is reached, in the case of the white dwarfs, by means of what is termed degenerate electron pressure. In effect, atomic nuclei are able to float contentedly in a sea of tightly packed electrons— as long as the star's gravitation does not exceed a critical value. By calculation, this figure is equivalent to some 1.4 solar masses, and is known as the Chandrasekhar limit. Degenerate stars heavier than this cannot resist the intense force of gravitation, which will serve to crush matter to still greater densities.

The next stage of compression beyond a white dwarf is that of a neutron star, which is formed when the electron sea is forced inside proton nuclei— producing what is tantamount to a super atomic particle comprised of neutrons. But just as white dwarf stars have their restrictions with regard to mass, so do neutron stars. In fact, it turns out to be a rather narrow limit, with the degenerate neutron pressure stage incapable of restraining gravity much beyond a few solar masses.

The question now arises as to the ultimate fate of stars with masses in excess of what can be supported by this last resort of degenerate neutron pressure. As bizarre as it may seem, the relentless force of gravitation will cause contraction to proceed with increasing efficiency— literally squashing the star to the point where light is unable to escape. An implosion of this magnitude will result in negation of normal space and time— *a stellar black hole.* (Detectable only through its gravitational effect upon other stars, a number of such celestial objects have been discovered and fully substantiate theoretical prediction.) Essentially, a black hole may be described as an entity that has severed many connections with the empirical universe and has entered *an entirely different realm of cosmic existence!*

Along with several hundred billion starry companions, our Sun resides in a vast aggregation of stars in a galaxy known as the Milky Way. In general shape, our system bears some resemblance to a fried egg, in that its central nucleus is thicker and more dense than its outlying regions, which tend to trail away at the edges. Through a study of 21-cm radiation, radio telescopes reveal a spiral structure that is so common among other galactic systems. With a thickness about one-tenth of its 100,000 light-year diameter, the lens-shaped main body of our galaxy is enveloped by a spherical halo of more sparsely distributed stars.

Just like the discovery that our Earth is not at the center of the universe, so astronomers have determined that the Sun is not located within the central bulge of the Milky Way, but is instead situated between two spiral arms roughly 30,000 light-years from the galactic nucleus. With an orbital motion of close to 150 miles per second, it takes our Sun almost 250 million years to complete one full circuit around the center of our galaxy.

The enrichment of interstellar gas clouds, through dust from supernovae explosions, has produced two basic types of stellar population: metal-poor and metal-rich stars. The former constitute the oldest stars in our Milky Way (having been born in a primeval environment consisting of almost pure hydrogen, with perhaps a trace of helium) and are believed to have ages of roughly 15 billion years. The second group of stars are all later creations and contain a much higher proportion of nature's heavier elements. Situated near the galactic plane, our Sun is a metal-rich star due to its later birth in a gaseous nebula enriched from earlier generations of supernovae. In contrast, the spherical halo of stars is populated by a preponderance of older metal-poor inhabitants, by reason of a lack of enriched gas clouds at the time of their formation.

Although optical telescopes are unable to penetrate the opaque dust clouds which hide the nucleus of our galaxy, this is fortunately not the case with regard to radio observation. The picture to emerge is one of interest. An extremely powerful radio source, known as Sagittarius A, is located at the center of the Milky Way. Coming from a region only about 40 light-years across, it is a strong emitter of synchrotron radiation— being produced by high-speed electrons spiraling around an intense magnetic field. By far the most logical explanation involves the presence of a *massive rotating black hole* at the center of our galaxy, possibly of the order of some 5 million solar masses if fast-moving gases near the galactic core is any indication. (Subsequent study of many other galaxies confirms the premise of powerful central black holes, which appear to be a common galactic feature.)

Forming a concentric halo about the nucleus of our galaxy, at distances of some tens of thousands of light-years, are several hundred orbiting globular clusters. Typically about a hundred light-years or so in diameter, and containing 100,000 suns (or more), they are exceedingly compact aggregations of metal-poor, older first generation stars. Featuring many pulsating stars, and with little uncondensed gas with which to form new stars, their common origin and distance made it easy for astronomers to note a correlation of intrinsic luminosity with their pulsation period. Also, their distribution about the galactic nucleus served to give man a clear indication of his eccentric location in the Milky Way.

Situated in the Southern Hemisphere, and appearing as two fuzzy patches of light to the naked eye, are twin satellite systems known as the Magellanic Clouds. Classified as small irregular galaxies, both are comprised of many millions of stars and possess a mixed population with regard to age and chemical composition. The Large Magellanic Cloud is about 160,000 light-years distant; while the Small Magellanic

Cloud is slightly further away at a distance of some 190,000 light-years. Like swarms of globular clusters, they give the impression of being remnants left over from the formation of our Milky Way.

In the constellation of Andromeda, at a distance of a little over 2 million light-years (MLY), lies the beautiful spiral galaxy known as M31. (See front cover.) This great star system is, in many respects, a twin to our own galaxy. Similar in both size and shape to the Milky Way, it is encircled by a comparable number of globular clusters. It also possesses a like number of highly luminous Cepheid variables in its spiral arms. Along with about a dozen and a half smaller systems, mostly of the dwarf elliptical type, this small clumping of galaxies comprise what is known as the *Local Group*— a relatively minor gravitationally bound cluster a few MLY across.

Moving outward to a distance of about 12 MLY, we come to another small cluster which is dominated by the spiral galaxy M81. Located in the constellation of Ursa Major, this system could also qualify as a twin to our Milky Way. Other prominent members of this group include NGC 2403, a small spiral system, and the irregular galaxy M82. In the constellation of Canes Venatici, the much photographed "Whirlpool" galaxy is observed as a major component of yet another weakly bound gravitational association. Also known as M51, this system measures about 60,000 light-years across and is a prominent member of another cluster of galaxies some 30 MLY distant.

A very noticeable departure takes place some 50 to 60 MLY away, in the direction of the constellation Virgo. For instead of sparse aggregations of galaxies, only a few of which contain really large star systems, an extremely rich cluster is finally encountered. Centered by the supergiant elliptical (almost spherical) galaxy called M87, the sprawling Virgo Cloud contains many substantial members and is the closest example of the dense swarms which exist in the universe. (Truly a monster among galaxies, M87 is no ordinary star system. Its turbulent central core is believed to contain a supermassive black hole of several billion solar masses!) This giant Virgo clumping forms the nucleus of a vast assemblage of lesser clusters— including our own

Local Group— and is known as the *Local Supercluster*. Similarly, as the Sun is not centrally located, so our Milky Way is revealed to lie at the outskirts of this overall aggregation.

Typically, a supercluster contains several thousand members and tends to be spread out over a spherical zone of perhaps 150 MLY. With an average separation between supercluster centers focusing upon a figure of roughly half a billion or so light-years, immense voids are seen to be interspersed with such clusters. Extending, upon occasion, to distances of at least 300 MLY across, one of these seemingly empty voids could accommodate many thousands of Milky Way galaxies in terms of Local Group separation. A subject of growing interest among astronomers, superclusters appear to constitute the ultimate degree of structuring in the universe.

The Expanding Universe

The principle of the Doppler effect is crucial to an understanding of what must be considered a truly fundamental feature of the universe. Just as with sound waves, radiation from an approaching source will result in a bunching-up of signals and a shift to shorter wavelengths. Conversely, a receding source will cause radiation to be spread out more thinly and shifted to longer wavelengths. Moreover, for practical purposes, the degree of this displacement is essentially in proportion to the speed of relative motion— in the sense that the greater the velocity the greater the wavelength shift. Since the light of a luminous object may be readily broken down to various reference points, by means of passing it through the prism of a spectroscopic device, it became feasible to determine speeds of approach or recession by measuring the amount of such displacement from normal or "rest wavelengths." In this manner, scientists were able to show that a shift in wavelength, toward the blue end of the spectrum, was indicative of a certain speed of approach; while a shift in the opposite direction, toward the longer wavelength of red light, was indication of a specific motion of recession.

The procedure soon became a powerful tool with which to deduce a wide variety of celestial facts, ranging from the revolutions of stellar binary systems to the rotations and movements of entire galaxies of stars. Utilizing this spectroscopic technique, the astronomer Vesto Slipher (Lowell Observatory) came up with a rather amazing discovery by the year 1925. Inexplicably, some 38 out of the 40 spiral nebulae that he had investigated showed redshifts, which was promptly interpreted as motion of recession. Furthermore, he measured speeds as great as 2% of the velocity of light— an incredibly high degree of motion in those days.

The full significance of this peculiarity was not appreciated until 1929, when another astronomer by the name of Edwin Hubble supplied the missing clue. Having just recently proven that the heretofore mysterious "spiral nebulae" were really distant galaxies of stars, he noticed a most remarkable relationship when the velocities of redshifted galaxies were compared with their distances. Without a doubt, the more remote systems were receding at greater speeds than nearby galaxies. Once distances much beyond our Local Group were examined, the entire universe appeared to be in a state of expansion! That is to say, except for relatively close gravitationally bound clusters, the galaxies were all receding from each other— and with speeds evidently in direct proportion to distance. Thus a cluster situated at twice the distance of another was deemed to be rushing away at twice the velocity. Likewise, one at triple the distance was seen to be receding at triple the speed, etc. Known as *Hubble's constant*, this rate of expansion of the universe may be defined as a measure of the time required for a receding system to double its velocity of recession. In terms of cosmological and philosophical implication, it was to be one of the most important revelations of the century.

Applying this remarkable distance/velocity relationship (generally referred to as the *Hubble Law*) to galaxies so remote as to preclude distance determination by other reliable means, it has been possible to map the brighter cluster members out to distances involving billions of light-years.

For some decades astrophysicists have been inclined to denote the degree of redshift by the letter "Z." For low velocities Z is roughly equal to the amount of spectral displacement expressed as a similar fraction of the speed of light. An object redshifted by 5% would be said to possess a redshift of Z=0.05; while one displaced 10% would be rated as Z=0.10, etc. This simple straightforward relationship begins to run into serious problems, however, with the contemplation of extreme velocities. In fact, as redshifts of higher and higher values were discovered, it was soon wondered just how long observers could go on breaking previous records. Eventually, spectral shifts were indeed measured which implied velocities of recession well in excess of the speed of light! Realizing the flat impossibility of such a scenario, in terms of traditional physics and established views of radiation propagation, it became necessary for science to introduce so-called "relativistic adjustments" in order to reduce velocities of recession to more plausible levels.

The mathematical form of this adjustment, which will assure a final figure that is less than light speed, may be stated as follows:

$$1+Z=\frac{\sqrt{C+V}}{C-V}$$

Where:

C = velocity of light
V = velocity of recession

Thus a receding source with Z = 0.24 will have its spectral lines displaced by some 24%; while Z = 1.00 will be equivalent to 100%; Z = 2.00 will reflect a shift equal to 200%, etc. Upon the basis of formula, such examples will be seen to have adjusted velocities of recession of some 21.5%, 60% and 80%, respectively, of the speed of light. (This relationship, between Z and velocity of recession, is depicted rather precisely by the curve in Fig. #2, where it really serves to express the *time of light transit* in full compliance with the dictates of an expanding universe and the "law of addition of speeds.")

In making this forced adjustment of speed and distance, there is now reason to suspect that science has actually managed to misuse and distort the true function of a perfectly valid formula. Instead of looking to Einstein's celebrated theory of relativity for an explanation (and entertaining only partially conceived ideas of increasing mass and slowing of time with rapid motion), **this formula is none other than a surreptitious reinstatement of the "law of addition of speeds"** — a fundamental and valid law of nature which was supposedly (and quite unjustifiably) abandoned toward the end of the 19th century, as a direct result of an unfortunate misinterpretation of the Michelson-Morley light experiment! (Of truly immense importance to a proper understanding of a host of cosmological mysteries, this vital issue of radiation propagation will be discussed at length in chapter 3.)

Quasars

Few discoveries by the science of astronomy inspired more initial controversy than detection of the first quasars back in the early 1960s. Characterized by a range of unprecedented extreme redshifts, these peculiar starlike objects immediately posed both an enigma and a challenge for man to use them as distance indicators in his quest to map the far reaches of the universe.

The initial mystery hinged upon the realization that quasars are incredibly luminous for their size. If these extreme redshifts were truly indicative of great distances— in the sense that they implied enormous recession velocities due to expansion of the universe— then their prodigious output of energy seemed incomprehensible in the light of contemporary physics. How could the energy-emitting region of a quasar, perhaps no more than a few light-days across, possibly liberate up to a hundred times (or more) as much energy as a giant galaxy containing hundreds of billions of stars?

Within a decade or two it was generally agreed that a supermassive black hole must somehow be involved if the inferred energy generation is indeed real. Rotating with extreme rapidity, the tremendous gravitational energy of a massive black hole is greatly supplemented by the dynamo effect of a powerful electric field sufficient to draw vast swarms of encircling quanta into intimate association. The outcome of such a dense concentration of swirling energy forms, according to theory, is a perpetual deluge of newly-created electrons and positrons spewing from the vicinity of a quasar's black hole nucleus. In any event, it is admittedly more than a little ironic that a supermassive black hole, from which nothing is expected to escape, should be held responsible for powering the most luminous objects in the universe.

Over the years evidence has continued to mount in support of a "cosmological" interpretation, which now appears to be firmly established in the minds of astronomers. Velocity of recession is, first of all, a known cause of redshifts. Second, the finding of some quasars with spectral line displacements similar to those of certain surrounding galaxies— in apparent compliance with Hubble law— is strongly suggestive of a common origin and location in space. A third indication arises from the fact that there are no quasars whatsoever within our own supercluster— a clear hint that separations well in excess of several hundred MLY are involved.

On the whole, it seems extremely unlikely that any reason will be found to negate the value of quasars as distance indicators. Typically a hundred times (or more) as luminous as a giant galaxy, with some modification of radiation propagation theory there is now every promise that these exotic objects will afford the necessary means with which man may probe to the very edge of the universe.

3

The Nature of Gravitation

Gravitational Attraction — Gravitational Repulsion — Propagation of Radiation

It is true, of course, that science has made considerable progress in uncovering the basic laws whereby gravitation exerts its influence, and also with respect to noting the curious relationship which exists between mass and energy. But in spite of the many experiences imparted to us by the force we call gravitation, it is evident that this knowledge concerns only effects — *the actual root Cause or Motivation itself remains very much unanswered!* Why, for instance, should any particle of matter have an attraction for another particle? Why does the smallest amount of material contain an astonishing quantity of what might be termed "congealed energy" or "captured motion?" What, in fact, does energy or motion really signify? Proceeding upon the basis of spiritual theory, there is reason to suspect that such issues are now capable of a considerable degree of clarification. Indeed, it may well be that many mysteries in the realm of physics will yield to just such an interpretation.

Gravitational Attraction

Recalling that the very desire and destiny of all spirit — from the instant of its conception as a lowly "**A**" — is ultimate fusion into the One Undivided Body of God, need we be surprised to observe that a small spark of this inherent desire is manifested by every particle of matter? Essentially, what we choose to call the force of gravitation is none other than an expression of the *desire for spiritual advancement* which prevails at the more elementary levels of creation — of the constant striving to overcome an inferior and divided state by means of a literal fusion into fewer entities of higher value. Were we to seek confirmation of the principle of spiritual fusion, from the physical world which surrounds us, *no more graphic illustration could be found than that posed by the phenomenon of gravitation!*

In order to separate unworthy and imperfect levels of spirit, and so prevent an unjustified fusion into One Harmonious Whole, nature has seen fit to introduce the property of distance. Such attributes of the universe as motion and energy are directly related and may, in fact, be looked upon as still another indication of imperfection — *or absence of Total Communion.* It also follows that *rate of communion* is a factor to be considered in conjunction with distance and relative motion. Should we theorize interaction among components of creation to be instantaneous, for instance, it would be equivalent to the elimination of both distance and time. Speed of communion is thus seen to play a vital role in the overall scheme of cosmic events.

Conceding that the phenomenon of gravitational attraction, among the various material particles, is actually a manifestation of such spirit's *desire for communion* (or fusion), the question of propagation becomes of paramount importance. Is this influence restricted to some finite velocity? Or, on the other hand, is it instantaneous in effect, and therefore immune to all events which might otherwise be associated with time? Although obviously propagated at an extremely rapid speed, both theoretical and observational evidence appear most implicit in substantiating the view that gravitation is not instantaneous. Of all conceivable suggestions as to the velocity of gravitational communion, it would seem that the answer must center around the *speed of light* — clearly a most significant and universal constant of nature.

With this thought in mind, and acknowledging the presence of vast swarms of unseen entities at the lower extremities of the evolutionary pyramid, we may now ask how the influence of gravitation is imposed upon bodies that are both relatively distant and massive. By what specific mechanism are particles of matter able to attract each other? Can we somehow utilize this impressive energy source which is presumed to exist in the form of countless minute quanta? While the basic concept of gravitational quanta — or infinitesimal bundles of energy — being exchanged by material bodies is not altogether new, the mechanics of how this interaction is converted into a means of attraction has hitherto defied explanation.

A vital clue to this ancient riddle very probably resides in what has been termed Newton's *Third Law of Motion:* "For every action there is an equal and opposite reaction." In short, what is implied in the case of gravitational attraction is that, whatever communion a body may receive from a source outside of itself, so it must promptly emit or relinquish an equivalence of such energy. (Were a material particle able to retain every quantum it encounters it would, of course, tend to grow infinitely greater.)

As a prelude to an attempt to resolve this problem, and giving due recognition to the principle of a constant exchange of energy among material structures, certain postulates may be stated as follows:

(1) Gravitational quanta differ fundamentally from such stable material particles as electrons and protons only in size or status and may be described as elementary forms of matter which have yet to evolve — through the process of fusion — into more advanced structures. Consequently, in compliance with the basic premise that relative abundance is a factor of intrinsic worth, a wide variation in the status or mass of such quanta is bound to exist. That is to say, the lower the status the more abundant the entity.

(2) Far from being instantaneous in effect, the propagation of such quanta is restricted to some finite velocity — essentially, a speed strongly focusing upon that of light itself. It may be presumed, in fact, that light speed represents *"escape velocity"* of infinitesimal quanta from association with some larger structure.

(3) In general, when a typical material particle encounters and absorbs a quantum of gravitational radiation it will be unable to retain this additional energy and must quickly emit any surplus. Moreover, unless prevented by some infrequent occurrence, there will exist a pronounced tendency to eject this surplus in the *opposite* direction from whence it was received. In this event, the inferred direction of emission is likely facilitated by the rapid rotation of the absorbing body. Unable to penetrate the dense concentration of halo energy which envelops a major material particle, a tiny gravitational projectile will be carried momentarily in the flow of this overall momentum. Finally, after rotation of the principle body has proceeded to the point where it no longer poses an obstruction, the newly acquired quantum is permitted to sever its association and resume its previous course with an escape velocity equal to that originally possessed — essentially, the speed of light.

(4) As infinitesimal quanta approach within a critical distance of a material particle, they will tend to come under the influence of an intangible force which might well be defined as anti-chance. The end result of this factor is to incite a greater influx of energy in the immediate vicinity of the material body, so that more quanta are received in unit time than could be explained on grounds of chance encounter alone. Such dynamic intervention may be ascribed, in principle, to the inherent desire for fusion which pervades all of creation. As a consequence of this inborn desire, every particle of matter may be visualized as possessing a *"halo"* in the form of an encircling cloud of energy — an atmosphere of unassimilated and relatively minute quanta which is prevented from obtaining an unconditional escape due to the constant striving of spirit to secure

fusion. Upon this basis, it will follow that halo characteristics must be determined somewhat by the mass of the central body, in the sense that a more massive nucleus will exhibit a more extensive halo and be capable of retaining larger quanta.

Proceeding upon the assumption that infinitesimal quanta are forever being absorbed and emitted by all material bodies, the task remains to relate gravitation with this exchange of energy. How is it possible to achieve attraction through innumerable encounters with these tiny quanta, acting as they must like a swarm of miniature projectiles?

Predictably, initial reaction to a larger than average flow of energy from one particular location in space — the direction of most dominant mass — will be a slight recoil away from the point of greatest impact, as numerous high-speed collisions serve to drive the body backward. Within the smallest fraction of a second, however, this motion of recession is terminated by the expulsion — in the *opposite* direction from whence received — of the excess energy just obtained. Acting upon the principle of jet or rocket propulsion, this ejection will quickly return the particle to its original position. At this point, the attracted mass will be made to oscillate rapidly with each cycle of absorption and emission, with the most conspicuous motion occurring along the plane of greatest interaction. The degree of this wavelike motion would be largely determined by the mass of the particle involved, in response to the logical premise that the more massive the body the less inclined it will be toward oscillation.

The key to the actual process, whereby small-scale oscillation is transformed into large-scale motion of attraction, would appear to have a solution in the production of what might be described as *"neutral quanta."* Just as swarms of infinitesimal projectiles may collide with bodies of much higher mass, so in regions of pronounced interaction numerous collisions must also occur among the tiny quanta themselves. Following a collision of two such quanta, there is bound to prevail a brief instant when their respective motions through space are retarded or neutralized to a significant extent, or even canceled altogether as in the case of head-on collisions among quanta of similar size. Considering that impacts of this type will be far more frequent in regions of concentrated mass, and especially along the plane of most intense interaction (or exchange of energy), it will be inferred that therein exists a substantial supply of "neutral quanta" which an attracted body may absorb and convert into a means of propulsion. That is to say, when incoming and outgoing quanta collide in the immediate vicinity of some larger body it may then become possible to temporarily absorb this "neutral" energy into such a system — *without penalty of incurring a backward thrust in proportion to the energy received* — and thence to utilize this potential as "fuel" by reason of its subsequent ejection at essentially the velocity of light. Thus the mechanism of gravitational attraction might well be explained through the acquisition of "neutral quanta" — *energy which may be absorbed into a system with reduced or negligible impact, but which is later ejected in the opposite direction at high velocity and with a recoil action sufficient to propel the particle in the direction of this source of energy!*

One important aspect of this proposal demands clarification. As an attracted particle is made to oscillate, through absorption and emission of energy, it will follow that the strongest concentration of potential "fuel" must tend to lie along a narrow plane extending from the attracting mass straight through the center of the attracted body. But since the combined flow of outgoing and incoming energy will be the same on opposite sides of an attracted particle, it may be deduced that the frequency of collisions will be such as to produce as much "neutral quanta" on the side opposite the attracting mass as on the reverse side. Obviously, at this point, a successful interpretation of acceleration must hinge upon the acquisition of a larger proportion of this "fuel" from the direction facing the attracting mass. Unless this can be achieved, the theory will collapse of its own accord.

Invariably, this preferred assimilation will be facilitated by the general flow of energy itself — a stream which flows most intensely in a direction from the attracting body toward the attracted particle. For one effect of this predominant stream must be to create a certain proportion of *secondary* collisions, as "neutral quanta" produced by earlier impacts are in turn struck by other members of this stream before they have had time to fully recover their momentum. Of fundamental importance, however, is the realization that when a secondary collision occurs between two interacting quanta the tendency will be to sweep this energy *toward* the attracted body. On the other hand, when such collisions are produced on the opposite side of the attracted particle, this potential source of "fuel" will be swept away by the force of the prevailing stream, since the direction of flow is away from the attracted member. As a result, a high percentage of any "neutral quanta" will be absorbed from the direction immediately facing the attracting mass. With the subsequent ejection of this "fuel" — in a direction *opposite* to that received — attracted bodies are thus enabled to propel themselves toward any influencing mass.

It now becomes desirable to assess certain properties of the halo (which is believed to surround all material particles), with particular emphasis upon the efficiency of the attraction process among particles of unequal mass. In short, will halo size and composition have any bearing upon the mechanics of gravitation? Both theoretical and experimental evidence would seem to leave little doubt in this respect, revealing that halo strength does indeed constitute a factor. Gravitational efficiency must surely vary with the mass of an attracted particle, to the degree that larger and more evolved structures, with their stronger and more extensive halos, will not respond quite so readily to acceleration as will smaller entities with more modest halos.

The explanation for this would appear to reside in the "insulating" properties of a halo. That is to say, a more extensive halo will serve to scatter or dissipate incoming quanta more readily than a smaller halo, thereby *insulating* the nuclear core from the "neutral quanta" that is so essential in achieving acceleration. For it is only when the thrust of ejected quanta can be imparted to the site of actual mass — such as the nuclear body — that genuine work may be accomplished in moving the overall structure.

Hence, it is to be noted that radiation in general is of a size that can easily be boosted to maximum velocity due to a favorable mass/halo relationship; while considerably heavier particles, such as electrons and protons, are more prone to squander energy and cannot be accelerated to the same degree. Protons, in particular, are noticeably more resistant to any attempt to push them toward light speed. In effect, due to the increased scattering of "neutral quanta" by a larger halo, only a diminished proportion of energy may be utilized by the nuclear body to secure acceleration.

Proceeding upon the aforementioned premises, it will be seen why an adjustment to Newtonian mechanics is required in order to express the factor of motion — simply because any application of external force must reflect the inability of gravitation to propagate at full efficiency when communion is limited to the velocity of light. However, rather than be mislead by traditional relativistic notions of increasing mass with extreme motion, it will be acknowledged that it is really a case of *declining efficiency of gravitation with increased relative motion*, as such movement makes it harder for quanta to catch up to rapidly receding bodies! Accordingly, at the speed of light itself the influence of gravitation must be virtually zero. The popular but highly nebulous idea of mass becoming infinite at light speed is both illogical and contradictory. (Indeed, radiation is propagated at just this speed and nobody has ever claimed infinite mass!) A modified relativistic formula, depicting a reduction of gravitational efficiency with increased relative motion, is illustrated in Fig. #4.

Gravitational Repulsion

In actuality, *not all infinitesimal quanta may be exchanged in such a manner as to induce attraction!* As opposed to traditional belief, gravitation can no longer be viewed as an attractive force propagated indefinitely into space — one which merely "varies directly as the product of mass and inversely as the square of the distance." For while this classical law is essentially valid within our Solar System, and may be applied with some measure of confidence to galactic phenomena in general, there is every reason to suspect that this is not the case when extremely large distances (such as exist between clusters of exterior systems) are considered.

According to theory, the susceptibility of infinitesimal quanta to the influence of material particles hinges somewhat upon the issue of size. In other words, smaller quanta will pose less of a challenge to a spiritual body attempting to exert its will — or desire for fusion — over other members of creation. (It is for this reason that the halos enveloping "maximum" particles, of the order of electrons and protons, are believed to be comprised of the more elementary levels of quanta — in the sense that larger bodies will possess halos of more evolved quanta.) Larger corpuscles must greatly increase the task of securing association and will, therefore, be far more likely to incur rejection. Nevertheless, even quanta too large to permit halo membership may be induced to follow trajectories of some curvature. (The more substantial the quanta the less affected will be their paths.) Applied to the overall structure of an entire galaxy, it means that the vast bulk of all exchanged energy is caused to remain in circulation within the confines of that system.

But while the majority of ejected energy is unable to escape into the depths of outer space, it is also clear that this restriction does not apply to the more massive quanta. In fact, it is most logical to assume that a significant proportion of quanta — residing below the spectrum of light and radio emission — is indeed able to pass freely into space. Very conceivably, it could be the manner in which these larger quanta — escaping from clusters of galaxies — interact with the material particles of other systems *which holds the key to the observed expansion of the universe!*

Inasmuch as quanta of greater mass will, through sheer force, be capable of penetrating deeper into the halos of "maximum" particles — and so encounter regions of higher density or energy content — the end result is very likely to be a prompt rejection and scattering of this surplus, particularly toward the direction from whence it came. In effect, there will arise a situation whereby spiritual bodies are suddenly called upon to "process" more energy than they are capable of handling. As an injustice of this nature becomes more extreme, so it could demand a more immediate expulsion of such foreign quanta — quite regardless of whether they have had sufficient time to be whirled around to the opposite side of the principal body. Hence, in response to an injustice perpetuated against both host and invading quantum alike, an incentive will exist for the foreign corpuscle to promptly sever all bonds and rebound much as a ball will bounce off a hard surface.

We can now see how an exchange of energy, among clusters of galaxies, could lead to mutual repulsion and an expanding universe. For in the process of exchanging quanta, of a certain mass range, two distinct instances of repulsion will be inferred as this impacting energy interacts with the major particles of such systems. In addition to the initial impact, which is a force directed away from source, there will be a secondary repulsive thrust as this undeserved quanta is swiftly rejected and expelled in the general direction from whence it came.

Since only a relatively small proportion of exchanged quanta will so interact as to induce repulsion, it follows that any inherent repulsive characteristics will be entirely masked by the predominant tendency to induce a state of "localized" attraction. On the whole, this universal exchange of energy may be viewed as the simultaneous existence of two opposing forces — with the force of attraction simply being the stronger at more moderate distances. With contemplation of the separations which are known to exist between clusters of galaxies, the hitherto dominant factor of repulsion is finally permitted to gain the upper hand, as the only quanta capable of being exchanged over such distances are the larger corpuscles which must act to induce repulsion.

This "dual" concept of gravitation, incidentally, seems to be just what is needed to resolve a curious mass discrepancy involving clusters of galaxies. Astronomers of the 20th century were puzzled to find large inconsistencies in calculating the total masses of such associations by two proven methods. One procedure, known as the "virial theorem" technique, depends on measuring the motions of member galaxies and relating this information with the principle that speed — within a gravitationally bound system — is a factor of mass.

Alternatively, by computing the individual masses of the star systems comprising a cluster, simple addition will yield a value. Invariably, a most disconcerting discrepancy arises upon a comparison of the two methods. It was found that the kinetic energy of the galaxies implied masses far beyond that revealed by a more detailed determination of separate masses. Moreover, it was discovered that the larger the cluster the greater this "surplus" motion which seemed to demand interpretation as excessive overall mass. This strange paradox has persisted in spite of much study — presumably serving notice that something may be going seriously awry when traditional views of gravitation are extrapolated to really large distances!

Recognition of a long-range repulsive factor, arising as a result of a universal exchange of minute quanta, affords a most plausible explanation of this dilemma. At distances which separate cluster members, the usual overwhelming aspect of gravitational attraction is greatly weakened. Accordingly, while still overshadowed, the inferred repulsive side of this energy exchange is permitted clear expression. There will exist, in fact, a pronounced tendency for the constituents of a cluster to push each other apart. Prevented from doing so by a superior binding force, this energy must reveal itself in terms of enhanced galactic motions. In short, it is translated into the "surplus" kinetic energy which appears to undermine any overall mass determination. It is surely no accident that the magnitude of this discrepancy, which is generally a factor of 10 to 20, reflects the size differential between the diameters of the galaxies and the actual distances separating cluster members.

Upon the basis of a finite physical universe — and only a universe of finite dimensions — this underlying repulsive influence will be permitted an opportunity to express itself in rather dramatic fashion. Since nothing will be held to exist beyond the outer boundary of the universe which would exert an inward pressure upon nearby clusters of galaxies, it may be deduced that the outermost clusters will be free to utilize this repulsive force to the fullest. Moreover, once started, any outward movement must be cumulative in effect, in the sense that a recession of the outlying systems will reduce any inward pressure upon the more central members — thereby allowing them to share this large-scale motion without any loss of efficiency. Thus the universe will be caused to expand at a constant rate, in full agreement with the Hubble distance/velocity relationship which holds that the speed of recession is a direct factor of distance; that is, a twofold increase in separation will be reflected in terms of a twofold acceleration in velocity of recession.

Strong evidence in support of this distance/repulsion (D/R) aspect of gravitation may be found in the "bubble-like" arrangement of galaxies that has now become apparent. [1] Contrary to widespread belief, which would consider gravitation to be solely an attractive force, most galaxies are so distributed as to suggest an opposing influence; for they tend to reside toward the edges of "shells" or "bubbles" — with near voids at their centers. *This is precisely the pattern to be expected should gravitation manifest a "dual" nature!* Indeed, this observation is quite incomprehensible tn terms of traditional cluster dynamics, which should feature central condensations of mass — not the reverse situation with such regions being almost devoid of galaxies!

Yet another indication of this "dual" nature of gravitation may well be found much closer to home — in fact, within the confines of our own Milky Way. In strict conformity to Newtonian laws of gravitation, the outer portions of galaxies should rotate much slower than their innermost regions which feature higher concentrations of mass. To the great surprise of many astronomers, it was discovered that the outlying sections of all spiral and elliptical galaxies rotate considerably faster than expected. Why? Unwilling to acknowledge a repulsive side to gravitation, they are compelled to postulate large quantities of unseen mass in the extremities of these galaxies. What form this "missing mass" might take has continued to remain a big mystery, in spite of an abundance of wild speculation. Hopefully, it will not be long before it is finally realized that the issue is really one of "missing physics," in which the repulsive side of gravitation is conceded to supply the missing rotational impetus.

This repulsive aspect of gravitation may well be associated with the formation (and preservation) of spiral structure in galaxies. Considering that the extremities of all elliptical systems take much longer to complete a revolution about a galactic center than do the innermost regions, it has been wondered how spiral arms could persist after a few orbits. The very fact that such structure is so common is a clear indication that some subtle force is at work upon a long-term basis. By far the most promising theory to be advanced is a density-wave hypothesis, in which rotating shock waves play a major role in compressing gas and dust into regions of new star formation — producing luminous spiral arms. An essential requirement of this scenario is a pattern whereby matter is caused to describe elliptical orbits about a galactic nucleus, with each successively smaller orbit offset, relative to outer orbits, in the same direction as the overall rotation of the system. A lingering problem, however, has been one of defining the actual mechanism responsible for this prescribed systematic alignment of orbits. (In essence, we are likely dealing with exactly the same set of circumstances responsible for the perihelion advance [2] of the planet Mercury.) Although upon a vastly larger scale, there is every reason to suspect that the existence of galactic spiral structure is an indirect manifestation of gravitational repulsion.

Upon due consideration of a possible formula with which to define this repulsive aspect of gravitation, it seems highly probable that the solution lies in ascribing a *twofold increase of repulsion for every fourfold increase of distance!* Radiating from the immediate vicinity of a proton nucleus, as a negative force some 10^{20} times weaker than the positive influence of gravitation, it may be presumed to increase with separation as stated — becoming equal to that of attraction at a distance of roughly 17 million light-years. What is so very remarkable about this distance/repulsion (D/R) factor (see Fig. #5) is the excellent agreement with observation of a host of celestial phenomena — including galactic cluster members and an expanding universe.

Propagation of Radiation

A number of revolutionary conclusions must arise as a result of our new interpretation of gravitation. The mysterious "wave" aspect of radiation, for instance, now promises to make some sense. We can readily understand why the various material particles exhibit wavelike characteristics, as they are caused to oscillate through periodic absorption and emission of minute quanta. But can we adopt this same principle to account for the rhythmic wave motion as it streaks through space? Considering such a phenomenon as the polarization of light, it is evident that a more subtle explanation is in order.

By far the most logical solution would seem to involve the view that radiation does not normally consist of solitary corpuscles. Just as many stars are members of double systems, so it may be expedient to visualize radiation as primarily *miniature binary systems* — essentially, two entities revolving about a common center of interaction, while moving through space as a unit. (Polarized light may be explained as swarms of minute binary systems possessing a specific and common orientation in space.) Due to this rotary motion, a quantum of radiation will be observed to display a spiral or "wave" trajectory, with the particular wavelength itself simply requiring interpretation as *an expression of the degree of separation of the binary components.* (It will be understood, of course, that the actual separation between such members is not synonymous with the distance from one wave crest to the next, as the latter attribute is really a measure of how frequently the components complete their mutual orbits. Since orbital speed decreases with either reduced mass or increased orbits, in full conformity to established laws of motion, it follows that more massive and energetic components will be closer together — signifying shorter wavelengths — and hence able to complete more orbits in unit time than more widely separated members of smaller mass.)

In view of the obvious simplicity of this concept, and considering its ability to clarify an otherwise baffling situation, it may be wondered why science ever hesitated to adopt this picture of the wave-particle duality of radiation. By far the most conspicuous objection is one which even threatens to ridicule the very suggestion (especially when we contemplate radiation of higher wavelengths), since it would appear to involve the lack of an attractive force sufficient to retain the structure of such weak systems. For example, when we consider that most binary components would have to be assigned separations exceedingly vast by nuclear standards, it may be asked why these associations are not immediately disrupted by much closer encounters with far more massive and influential particles.

Upon the basis of a limiting velocity on behalf of gravitational interaction (e.g., the speed of light), and with insight into how acceleration is accomplished, it will be seen that by the time full communion could be established with any nearby material particle a tiny bundle of radiation will have left that particular region of space and will, therefore, be largely unaffected. (Invariably, an exceptionally close encounter or a collision would be another matter.) Thus, by reason of a common motion through space which would permit a state of mutual communion to exist, a binary system of quite minimal gravitational attraction may prevail — *so long as this system possesses the velocity of light relative to any body which might tend to disturb it!*

Having introduced a proposal whereby material particles may achieve attraction, it now becomes necessary to explore laws regulating this acceleration. What, for instance, prevents the acceleration of a body beyond the velocity of light? As a matter of fact, when we talk in terms of velocity it may well be asked what is really meant. To what specific point is any motion relative? If the speed of light represents some limiting velocity, then it is at once evident that a body possessing this "maximum" velocity relative to one system will not possess this same speed with respect to all other moving systems. Two light photons, traveling in opposite directions, will obviously exceed the speed of light relative to each other. Likewise, clusters of galaxies clearly exist which display combined velocities of recession in excess of this speed. Hence, in a broad sense, the velocity of light can be surpassed, and any inferred restriction very much remains to be clarified.

Essentially, the speed of a body through the medium of space is regulated by its ability to acquire "neutral quanta" — with acceleration proceeding just as long as it is possible to obtain and utilize such "fuel." Should we theorize motion in excess of the speed of light, energy encountered in the direction of flight will strike with greater force than the recoil action with which it is later ejected — thereby constituting a braking action. On the other hand, if a body is moving at close to light speed, relative to the prevailing gravitational field, it is evident that only a minimal amount of quanta could ever manage to catch up to it from behind in order to produce "neutral quanta." It is for this reason that velocity becomes so critical at extreme speeds, and which is responsible for the self-terminating aspect of motion once the velocity of light is attained with respect to the prevailing gravitational influence.

In assessing all aspects of acceleration, it will be noted that any adjustment of speed is the product of both velocity and density of the medium through which passage is attempted. In a sense, a lengthy transit through a tenuous gravitational field is seen to be equivalent to a shorter encounter with an intense field. Nevertheless, it is most essential to understand that it is the actual velocity of a body which becomes of overwhelming importance at extreme speeds. For as light speed is approached, the chief effect of a stronger gravitational field will be to merely expedite any required adjustment of velocity — *an adjustment which the weakest field will surely facilitate, even if it should take a slightly longer period of time!*

Certain consequences of extreme motion warrant discussion. It may be inferred, first of all, that a body moving through a gravitational medium at close to the velocity of light will exhibit an increase of mass due to the greater volume of quanta swept up and momentarily detained, where it must add to the strength of a particle's halo. But rather than be misled by a misconception held by many scientists, which would visualize mass as being increased by an infinite amount once the velocity of light is attained by a material body — an obvious impossibility — it will simply be conceded that further acceleration is precluded by reason of the inability to secure propulsive "fuel."

Seemingly, a notable feature of rapid motion is the curious effect of time retardation — a phenomenon which is believed to have received a measure of verification through laboratory experiment, even if the mechanics involved have been far from understood. It has been found, for example, that the time required for certain unstable atomic nuclei to undergo a spontaneous disintegration is a factor of velocity. Having little influence at low speeds, as the velocity of light is approached this retardation effect becomes quite pronounced. It may be wondered how an internal instability problem could be affected by external motion. What happens, of course, is that increased speed leads to the acquisition of a more substantial halo — an encircling swarm of minute quanta which serves to insulate the nuclear body from the outside world. But since insulation works in both directions, it follows that there will exist a tendency to preserve whatever resides within the halo. In order for disrupted components to escape, they must first break through the enveloping atmosphere of the halo, a feat which will become somewhat more difficult should the strength of the halo be enhanced due to increased motion within a gravitational field.

Upon a philosophical basis, it may now be concluded that the velocity of light represents a typical expulsion from an undeserved level of fusion, or spiritual association — *motion which is relative to the entity it has just left!* It may be assumed, in fact, that one basic and universal constant of nature is the speed whereby all undeserved quanta are ejected from the vicinity of material particles. In essence, the "escape velocity" imparted to expelled quanta is an expression of *unworthiness*, of total separation with respect to the system with which it was momentarily and unjustly associated. Applied to the propagation of energy through space this means that *whatever field of influence a tiny bundle of radiation may enter, so it will be caused to travel with the velocity of light relative to that gravitational system!*

It becomes possible, at this point, to gain much needed insight into a mystery which was raised by the classical experiment pioneered by Michelson and Morley toward the end of the 19th century. Until this time it was firmly believed that the propagation of light would obey the same fundamental laws of motion as does the material bodies with which we are so familiar. For instance, if we expel a projectile from a moving object — in the direction of its motion — it will inherit this initial velocity in addition to any later thrust. This phenomenon is known as the "law of addition of speeds." It was therefore expected that light quanta, emitted from moving sources and striking other moving objects, would display variations in velocity commensurate with such movement.

The results of experiment, however, were most conclusive and left no doubt as to the fact that something very strange must transpire with respect to tiny photons moving with extreme velocity. It was found that radiation is propagated at what appears to be a "constant" velocity; for regardless of the motions of either source or receiver the speed of impact was the same! It was as if some intangible force had intervened and accelerated any photons which approached at less than a certain speed, and resisted or slowed the motion of those approaching in excess of this same limiting velocity. The nature of this mysterious intervention may now be defined, of course, in terms of absorption and emission of energy by radiation passing through the swarms of infinitesimal quanta so permeating space and constituting a gravitational field — essentially, a process which assures propagation at the speed of light relative to the prevailing influence.

Unfortunately, the Michelson-Morley experiment received a false and rather curious interpretation which is destined to go down in history as one of the greatest blunders of modern science. It was actually assumed that when a light photon is emitted in our direction, from a receding celestial object, it will at once begin to approach our system with an effective speed of some 186,000 miles per second (light speed) — while yet exhibiting an identical velocity of recession with regard to the object which it has just left. Initial relative motion — between source and receiver — is totally ignored! Accordingly, the distance of a receding source (in light-years) is also thought to be a measure of the time (in years) that has elapsed since this light first began its journey. But this view involves unwarranted and contradictory postulates. Not only does it necessitate an unexplained break with the "law of addition of speeds," but it relies upon sheer magic to be able to account for its miraculous ability to approach all bodies, situated in its line of flight, with a speed of 186,000 miles per second

— irrespective of whether such bodies are in fact approaching or receding!

This widely accepted interpretation is surely a grave mistake and, in many respects, a most bizarre and impossible concept. Far from ignoring relative movement between source and receiver it may well be inferred that, subject to comparatively insignificant time lapses as multiple minor adjustments of motion are quickly effected, *radiation is propagated through the depths of space with a speed commensurate with the dictates of an expanding universe, and in full compliance with the velocities by which such bodies are approaching or receding from each other!* For all practical purposes, radiation is subject to precisely the same basic "law of addition of speeds" as material particles in general.

In effect, without ever realizing the full implications of traditional redshift adjustments, astronomers have been using a calculation which actually takes into consideration the perfectly valid "law of addition of speeds" — a law which was supposedly discarded with the general acceptance of Einstein's relativity! Ironically, in the process of translating extreme redshifts, so as to derive lower and more reasonable values (see *The Expanding Universe* section in Chapter 2), they have inadvertently stumbled upon the very formula which is capable of describing the *time* of light propagation from distant receding sources. Instead of merely a means of defining a degree of spectral displacement, the adopted Z term is really *a measure of the time required for radiation to travel from source to detection within the confines of a finite expanding universe!* (See illustration, Fig. #2.)

All that is necessary to transform a Z redshift into the true time of light propagation is to multiply it by the radius of the so-called "observable universe," expressed in light-years — a figure which will always be some 1-1/2 times that of Hubble's constant. The reason for this, of course, is due to the fact that velocity (relative to the center of the universe) is accelerating steadily with distance. For example, at a point equidistant from the center and edge of a universe with an 18 billion light-year (BLY) radius, a source would be receding at 50% of the speed of light. In order to double both velocity and distance, in one cycle of Hubble's constant, it is necessary to travel a further 9 BLY and arrive at the edge of the universe with a speed equal to that of light. Since the average motion of such recession will be 75% of the velocity of light, it may be calculated that the time required will be of the order of 12 billion years. Hence, Hubble's constant will be 12 billion years in a universe having a radius of 18 BLY. (See illustration, Fig. #3.)

In actuality, the time of light transit from many distant receding sources will be enhanced somewhat by what might be viewed as a rather unique and hitherto unsuspected curvature of space. Due to the expansion of the universe and a steady outflowing of all matter forms — including the infinitesimal quanta so efficient in interacting with radiation — there must prevail a perpetual influence tending to push radiation toward the nearest edge of the universe. The end result of this outward streaming of space, will be the bending of radiation received from all directions not in harmony with this overall flow. In effect, *there will appear to be a general deflection of images toward the center of the universe* — a displacement which must include all extragalactic sources not in our line of sight with either the central core or the closest boundary of the universe!

The degree of this deduced "arcing" (or bending of the path of radiation), from remote sources partaking in the expansion of the universe, will hinge upon three factors:

(1) It is, first of all, an effect *which is enhanced by increased distance.*

(2) While quite incapable of imposing "arcing" upon radiation pursuing trajectories that are in harmony with the mass outward flowing of space, *it will be the most pronounced at right angles to this overall movement.* (See Fig. #3.)

(3) Such bending of radiation is strongly regulated by *proximity to the edge of the universe* — in the sense that extreme curvature must arise and become apparent to an observer situated close to the cosmic edge, where expansion velocity approaches the speed of light itself. (It is for this very reason that a host of celestial objects, of widely varying redshift, would appear to reside in close proximity.)

References:

[1] "A Slice of the Universe," March 1, 1986, *The Astrophysical Journal Letters.*

[2] *The Eternal Universe*, Harold W.G. Allen.
(ISBN: 09624555-1-2)

4

New Vistas in Cosmology

The Creation Impasse — Evolution of Matter — Conservation of Mass-Energy— The Distribution of Quasars—D/R Arcing of Radiation

With the advent of modern science a variety of cosmological theories have been proposed in an attempt to explain the origin and structure of our physical universe. We now see that such research has been flawed by incomplete and erroneous views of both gravitation and radiation propagation. In turn, this has led to a number of misconceptions and "blind alley" models which generations of future scientists may well regard with disdain. It would not be the first instance in which a little knowledge has served to provide the fuel for wild speculation; nor is it likely to be the last, as imagination is often the key to opening new horizons. What does invite criticism, however, is the blatant attitude of those professionals who refuse to consider alternate ideas and the possibility of past error. Contrary to general belief, and in spite of a profusion of scientific wonders, there is little freedom in certain branches of science relating to cosmology.

The Creation Impasse

For almost half a century the most widely accepted explanation for the origin of the universe has been a cosmology known as the *Big Bang* creation model. Inspired by the knowledge that clusters of galaxies are all rushing away from each other, it was perhaps inevitable that some highly imaginative scientist would theorize the universe as the product of a gigantic explosion. Moving outward from a common center, it was first assumed that all of the elements were formed within the primordial fireball shortly after the momentous event of creation. This hypothesis was soon modified to exclude the formation of atoms heavier than hydrogen and helium, when it was realized that these elements are produced within the interiors of massive stars. (Also playing a role in this revision was a technical difficulty involving the build-up of complex nuclei under the prescribed conditions.) According to this concept the impetus of an explosion, which is thought by many to have occurred between 10 to 20 billion years ago, became translated into an expanding universe— from which clouds of hydrogen atoms later condensed into stars and galaxies.

Certain philosophical implications are inherent in the nature of this view. First and foremost, it implies that the universe was created at one specific moment in the past. Hence, at some time in the future, it will have aged to the stage where it is no longer able to support biological life. Upon this premise clusters of galaxies will simply continue to recede from each other until, finally, the sky will be void of all exterior systems beyond one's own gravitationally bound cluster. This model is finite in both time and physical constitution. An obvious issue which arises involves the *Purpose* of creation. What motivation lies behind the origin of matter, and of the subsequent life forms which it nurtures? On the basis of a Big Bang cosmology the universe appears highly transitory in essence. Moreover, if God be the Creator, where was He before? Did He also come into existence at the same time as the universe? More questions are surely raised than are answered.

From a purely scientific standpoint such an explosive origin is beset by serious inconsistencies involving the *time* factor. In brief, the ages of some celestial objects do not appear compatible with the supposition that the universe was created at one particular instant in the past. While certain portions are unquestionably very old, other regions give every indication of comparatively recent formation.

To the consternation of many Big Bang proponents, a *Johns Hopkins News Release* (July 7, 2004) clearly reveals a rather disconcerting picture of galactic evolution. It is reported that, in a survey of some 300 high-redshifted galaxies, a surprising number of massive and fully mature galaxies have been found to exist only a few billion years after a supposed beginning of the universe! This unexpected evidence is totally against traditional cosmological viewpoint and evolutionary theory, which invariably features the formation of only small and immature galaxies at such high redshifts. As stated in the news release, astronomers have now discovered "highly developed galaxies that just shouldn't have been there, but are." (Needless to say, this observation is compatible with dictates of the "New Cosmology," which predicts a mixture of massive/older galaxy clusters interspersed with numerous smaller/younger galactic systems.)

Building upon certain supposed advantages inherent in a Big Bang cosmology, a modified oscillating version was soon developed and came to be favored by some cosmologists. This concept visualizes recession of the galaxies as being slowed by gravitation. Upon this premise it is inferred that expansion will eventually be converted to contraction, so that all matter may be brought together with such force as to produce periodic explosions on a truly cosmic scale. In effect, the universe would be seen to oscillate between states of expansion and contraction. Although appearing to eliminate the problem of initial origin, there has been much debate as to whether or not sufficient mass exists (even upon the basis of traditional views of gravitation) with which to produce a closed universe. But a crucial objection to this model is posed by the thought of radiation escaping from the outer extremities of such a universe; for there is simply no way for lost energy to be recovered! To appeal to gravitation is not only useless, but would imply that the weakest field could exert a drastic influence upon the motion of radiation — in itself, a supposition directly opposed to traditional views of light propagation! Should just one quantum escape, with each cycle of expansion and contraction, in time an oscillating universe would be literally dissipated into space. Thus the concept has all the aspects of a monstrous contradiction.

In a last desperate attempt to save the Big Bang principle yet another wild modification was proposed: *the inflationary Big Bang.* After a closer look at the traditional Big Bang model had revealed a host of blatant contradictions with observation, it became quite clear that something drastic had to be done. Not only did a remarkably smooth microwave background fail to agree with the large-scale clumping and distribution of galaxies, but the rate of expansion of the universe was deemed insufficient to solve a variety of additional problems which had surfaced over the years. Thus it was theorized that, at the precise instant of the supposed explosion, the initial expansion rate was such that it would actually double in size every 10^{-35} seconds! Following a hundred of these alleged doublings, it was rashly presumed that the universe would, in rather magical fashion, then revert to the rate that is presently observed! Needless to say, this concocted (but widely acclaimed) scenario reminds one of earlier attempts to preserve an incorrect and outworn picture of our Solar System — one in which the Earth was viewed as the central member.

Incidentally, it is to be mentioned that the role that Einstein played in supporting an explosive origin of the universe is most curious. It is rather ironic that he chose to retract his "cosmological constant" after the Big Bang model became so popular. (This factor — featuring a "repulsive" side to gravitation — was a later addition to his field equations when it was pointed out that the universe could not remain static, but must either expand or contract if gravitation is propagated indefinitely into space as an attractive force.) In so doing, he probably made the greatest mistake of his entire career. Had he not embraced the Big Bang principle, it is doubtful if so many scientists would have been persuaded to follow in his footsteps. In what is surely destined to be recognized as a travesty of science, this blunder by Einstein will be seen to convey a valuable lesson. It is to be hoped that future generations of scientists will soon learn to trust their own faculties of reasoning, instead of being so unduly influenced by the prestige of an esteemed colleague.

An interesting alternative, known as the *Continuous Creation* or *Steady-state* principle, was proposed shortly after the original Big Bang cosmology. This model held that the universe is infinite in time, in the sense that as matter is lost through the process of stellar evolution and cosmic expansion it will be replaced by the creation of an equivalence of new matter in the voids between galaxies, so that the overall nature and structure of the universe is preserved. However, the inability to account for the powerful driving force behind cosmic expansion was promptly decried by critics to be an exceedingly weak feature of this model. Discovery of the microwave background radiation (only later to be subjected to valid criticism) was at once declared to be the fading remnant of a fireball announcing the birth of our universe. During the 1960s yet another blow to this model appeared to be delivered by the radio telescope. It was found that the number of fainter radio sources was greater than what might be expected from a simple increase of volume with distance. Since, in looking out into space one is also going back into time, it was inferred that — upon the basis of an explosive origin — man should be able to see the more distant portions of the universe when objects were less scattered through expansion, in apparent agreement with observation.

Now the astonishing conclusion to emerge is that the scientific evidence, as currently interpreted, stands in flat contradiction to all of these cosmological models in some crucial respect. We know that this cannot be so, since one of the two basic creation choices must be essentially correct. (Either the universe always existed or it was created at some point in the past.) Nevertheless, this is exactly the prevailing state of confusion and frustration posed by the study of radiation from distant sources — information enabling man to probe billions of years into the past. The present overwhelming preference of astronomers for the Big Bang principle is due, not so much to the absence of criticism, as it is to seemingly insurmountable objections to the alternate Continuous Creation cosmology.

As a result of this curious impasse, which is reached when one attempts to determine the true cosmological model, it is well to ask whether this dilemma might not arise because of an erroneous assumption with regard to the manner in which radiation is propagated within an expanding universe. Indeed, it is surely more than a mere coincidence that, in almost every instance, the element of *time* is somehow directly involved in a damaging contradiction of the various creation models!

What is so promising about our study of radiation propagation is that redshifts, detected in the light received from distant receding systems, contains a hitherto unrecognized factor. Such Doppler shift measurements must reflect not only genuine velocity of recession, but also the distance — expressed in terms of an apparent increase of velocity — by which the universe has expanded during the time that has elapsed while radiation travels from source to detection. In short, the redshift in the light of a receding system is really a composite of velocity of recession and cosmic expansion — *essentially, it is a measure of the time that radiation has spent in transit!*

While it may be a valid premise to state that the velocity of light is a constant relative to the prevailing gravitational field, it will be understood that *such fields are themselves in motion* as they share in the expansion of the universe! Of foremost concern is the realization that any attempt to deduce the age and origin of the universe, from its rate of expansion, is likely to be fraught with unsuspected complications. Ignoring the valid "law of addition of speeds," it has been the custom of astronomers to accept the distance (in light-years) of a remote celestial object as also being synonymous with the number of years required by light to make this journey. We now see that this view must be incorrect and could readily lead to a gross underestimation of the time involved, since light emitted from a receding source is able to approach our system with an effective speed of 186,000 miles per second only after it has actually entered our own gravitational field. *By far the greatest proportion of its journey is characterized by a velocity of approach that is somewhat less than this speed!* Hence, there will be a discrepancy in time of light propagation, as implied by orthodox and revised concepts — a difference that is almost negligible for neighboring clusters of galaxies, but which assumes overwhelming importance when the more remote clusters are considered. (See Fig. #2.)

By virtue of this revised view of radiation propagation, it at once becomes possible to refute evidence which many astronomers had accepted as conclusive in eliminating the principle of Continuous Creation. Hitherto, the discovery of an excessive number of faint extragalactic radio sources (beyond what might be expected from a simple increase of space/volume with distance) seemed to support the Big Bang model. However, we now see that — at extreme redshifts — we are actually overcoming *more time than distance!* Rather than detecting larger numbers of remote sources in a smaller volume of space, it is seen to be a case of observing more objects of a significant age group. In effect, *we are able to view substantially more than one generation of creation!*

Should we accept this proposed revision of radiation propagation, which implies that we have already detected many objects with light transit times some tens of billions of years greater than what is permissible upon the basis of any explosive origin of the universe, *then there is no alternative but to conclude that the highly dynamic principle of Continuous Creation emerges in sole agreement with the evidence!*

What were once considered impressive objections to Steady-state cosmology are thus revealed to be without foundation. But the workings of prejudice are not readily overcome. An entire generation of astronomers and physicists have been taught to accept — without question — the idea of a Big Bang creation. It will not be easy for many scientists to admit that they have dismissed an alternative which should have been given more serious consideration. It becomes even more difficult when it is realized that this decision was based upon fundamental scientific precepts which are so obviously flawed!

Although it is almost inconceivable that present views of radiation propagation ever came to be accepted, this digression must be dealt with in the manner of genuine science — namely, through impartial analysis of all available evidence. In so doing, a truly amazing picture of the cosmos will be seen to unfold.

Nevertheless, it is also evident that Steady-state cosmology, as presently understood, is very much in need of certain elaboration and modification. In particular, it must be clear that creation of elementary matter forms cannot possibly occur at the level of the hydrogen atom (as originally proposed), which is now seen to represent a highly evolved state from the initial status of a lowly infinitesimal quantum, or "**A**." Also requiring clarification is the microwave background radiation, which is thought to argue in favor of an explosion from a superdense singularity. Similarly, many fundamental problems relating to the unique physical properties of matter remain unanswered.

Evolution of Matter

Without a doubt, a most pressing unsolved characteristic of matter is that of electricity, a strange influence which may be expressed as a force of either attraction or repulsion. While an influence of extremely short-range, it is an incredibly powerful phenomenon which completely dominates the nature of matter at close distance. But just what is electricity? Why do two "positive" or two "negative" charges repel each other; and why do two unlike charges attract?

A possible explanation suggests itself. Recalling the premise in which an inborn desire for fusion will induce a cloud of minute quanta to congregate about a material particle, it may well be asked if *halo rotation* might not afford a solution. What, in fact, are the consequences of postulating the existence of a cosmic law which recognizes but two fundamental and opposed states of particle rotation? May we not assume that "**A**" is created of opposite "charges" — essentially, *opposed* directions of rotation? Most assuredly, this assumption is in full agreement with the established scientific principle wich states: "For every action there is an equal and opposite reaction."

But how is it possible for charged particles to always display one specific direction of rotation (or charge) relative to other charged bodies — especially since they may approach at any one of an innumerable number of angles? Actually, such a problem is not nearly so formidable as it might appear, and would be seen to arise solely because we are inclined to visualize rotation as something which transpires in but one direction at a time. But any singular motion is capable of expressing only a flat two-dimensional universe, whereas it is clear that we live in a world of three basic geometrical dimensions. The difficulty is at once removed if, in addition to equatorial rotation, we also postulate an equal and simultaneous polar rotation. In other words, whatever polar and equatorial rotations may characterize particles of negative charge, so positive charges are to be interpreted as flows of halo energy in precisely the opposite rotary directions. Upon this premise it will follow that both like and unlike electrical charges will always exhibit properties consistent with their relative flows of energy, quite regardless of the angle of approach.

This entire effect, incidentally, may be demonstrated simply by bringing together two spherical objects, and by imparting the appropriate rotations in order to simulate a variety of like and unlike charges. It will be found that, no matter what angle the two bodies of *like* rotation are caused to come into contact, the motions of their respective halos will oppose or clash. Conversely, the two bodies of *unlike* rotation will always flow in harmony with each other at the point where they touch. It would therefore seem prudent to infer a connection with the effects of repulsion and attraction, respectively. (The various types of particle spin theorized by physicists, in an attempt to explain certain phenomena, may be reconciled with such a concept on the grounds that they merely constitute *secondary* motions superimposed upon a much faster and more basic trait.)

When two material particles of like charge approach to within a critical distance of each other their swirling halos will collide, with the result that a substantial proportion of their total energy content will be transformed into "neutral quanta" at the point of impact. Unless the nuclei of the two particles come together with sufficient force to burst through the encircling atmospheres of their respective halos — thereby accomplishing an act of fusion — this new concentration of energy must be released with explosive violence following a very brief period of compression. In essence, the repulsion of two particles of like charge may be attributed to a sudden exposure to more energy than can be "processed" (e.g., absorbed and emitted) without posing a critical injustice to the involved nuclear bodies, in the sense that such spiritual entities will be intimately associated with far more "**A**" than they deserve. In this event, the magnitude of the injustice will prevent a strong influx of "neutral quanta" from lingering long enough to be whirled around the body of a major particle and ejected in the opposite direction from whence received. Instead, this excess energy will be promptly rejected and expelled in much the same direction from which it arrived, thus imparting a strong thrust away from the vicinity of the impact site.

In direct contrast to this explosive state, we have the case of two unlike charges coming into close association. Since the two flows of energy are essentially compatible, at the instant of contact, the production of "neutral quanta" — through halo collision — will not immediately exceed the ability of the two approaching bodies to "process" this energy. What now transpires is really tantamount to raising the question of what prevents an unlimited contraction of a positive-negative pair of particles. Initially, the mutual penetration of halos must serve to produce only a modest increase of "neutral quanta," with an appropriate acceleration in velocity of approach beyond that which could be expected from gravitation alone. With the incidence of collision among halo quanta increasing as penetration progresses, the two rotary flows of energy will intersect at angles which tend more and more to oppose each other. As a result, a rising proportion of this "fuel" will be encountered in such a manner as to induce a spiraling motion — specifically, a thrust imparted at somewhat of an angle to a line which would characterize a direct approach. Following a brief period of intense acceleration, as this increased supply of "fuel" is put to use, the stage will be reached when orbital velocity becomes sufficient to offset any attractive force. At this point a balance will be established and the pair will enter into a stable orbit, rotating about a common center and separated by a distance commensurate with the total mass of the system.

It would therefore appear that we are justified in upholding the principle of opposed motion as a key to an understanding of this basic trait of matter — a trait presumably initiated with the creation of "**A**" and perpetuated through the evolution of infinitesimal quanta to such ultimate particles as the electron and proton. Upon a philosophical plane, one might choose to interpret this electrical aspect of creation as a divided state somewhat analogous to our own sex-divided world — whereby an incentive exists for all "unbalanced" members to seek mutual balance and progress through union with oppositely "unbalanced" entities.

We may thus come to the conclusion that the mysterious phenomenon of electricity is but a unique manifestation of the force we call gravitation. In all probability, an electrical charge is simply a *highly condensed rotary flow of gravitational energy* — perhaps little wonder that past failure to explain the nature of this universal force should also coincide with the inability to define electricity.

As an approach to the actual construction of matter — from its inception as a lowly "**A**" — a number of relevant mysteries may be cited. What, for example, determines the sizes or masses of the various atomic particles? How, in fact, are they even formed at all? If radiation represents a variety of particles intermediate in size between an "**A**" and the level of an electron, then why are such corpuscles unable to engage in evolutionary matter-building activities? Why are there no antiprotons outside of extremely short-lived laboratory freaks? Why are free electrons so much more plentiful than their opposite counterpart, the positron, which is practically an unknown commodity as a free entity in our section of the universe? How can the properties of neutral particles be explained? Why are neutrons so unstable outside of an atomic nucleus? Furthermore, how do they accomplish the feat of binding together protons which would otherwise repel each other?

Proceeding upon the premise of a Steady-state cosmology, we

may now address the problem of describing the actual evolution of matter from its initial status as an infinitesimal "**A**."

It is clear, first of all, that new creation cannot occur at random throughout the expanse of the universe. Most assuredly, it is likely to take place in response to a significant *absence of matter* — a vacuum of such exacting proportions as to be found only between the clusters of galaxies themselves. Thus we may visualize creation as a phenomenon transpiring in the voids between clusters of galaxies, and moving outward at just the required rate to replace that matter which is lost through expansion of the universe. As a result, new Creation Zones will develop and be centered between existing aggregations of matter as soon as density and gravitational influence reaches a critical minimum — from whence creation will spread out and assume the form of expanding shells as it follows the recession of surrounding clusters of galaxies.

But if the introduction of newly created "**A**" is confined to regions of extreme vacuum, how do elementary forms ever manage to condense — in the time permitted — to a degree sufficient to produce such highly compacted structures as electrons and protons? Obviously, at some critical stage, sparsely distributed "**A**" must be transported to regions of much higher density — from whence the required fusion process is enabled to function with vastly increased efficiency. The problem is thus seen to be one of transporting widely scattered quanta to a common meeting point.

Inasmuch as gravitational attraction cannot be evoked during the earlier stages of condensation, we are left with the *repulsive* aspect of gravitation as the only viable alternative. Converging from all directions in space toward the center of an expanding Creation Zone, between neighboring superclusters of galaxies, this steady streaming of energy facilitates a chain reaction effect. In time, the result must be a veritable deluge of elementary matter forms toward a central point of sufficient density as to ensure fusion into such fundamental structures as electrons, positrons and protons . Since the energy required to condense "**A**" is equal to that needed to push an equivalence of mass apart, it is tantamount to yet another demonstration of

Newton's Third Law of Motion: "For every action there is always an equal and opposite reaction." Essentially, this exchange of "repulsive" energy serves a twofold purpose, *as the condensation of "A" is accomplished at the expense of producing an expanding universe!*

The next major problem concerns the question of particle size. For while it is obvious that a direct fusion of larger and larger quantities of elementary matter forms cannot continue indefinitely, it very much remains to be explained how nature has contrived to impose the observed limitations in the sizes of such stable and "maximum" particles as electrons and positrons, and the somewhat higher level of the proton.

According to a firmly established scientific law, the rotation of a spinning body will be enhanced with either a reduction of diameter or an increase in mass leading to a rise in density. Inasmuch as the fusion of greater quantities of mass-energy into one larger mass will lead to faster rotation of the accreting body, it could well follow that a solution to the problem of particle size is to be found therein. For the very act of increasing rotation — through the incorporation of additional mass — must produce more violent collisions with newly encountered quanta. Should the force of these impacts exceed a critical value this excess quanta will be accelerated to escape velocity in an extremely short span of time. Furthermore, as rotation approaches the speed of light, so the efficiency of gravitational communion will be greatly reduced — *becoming totally negated once this limiting velocity is attained!* The evolution of a "maximum" particle is thus seen to be a self-terminating process by reason of rising rotational velocity, which prevents assimilation of additional quanta.

Yet another factor which effectively limits the acquisition of quanta, by an evolving particle, concerns *time of communion,* as sufficient time must be given for a critical degree of communion to be established with newly encountered quanta. In short, relative velocities must be such as to allow *intimate association for a minimum interval of time!* Again, relative velocity is seen to be the key to an understanding of what limits particle size.

It will therefore be inferred that the evolution or development of material structures may proceed in a more or less straightforward manner until the electron-positron level is reached, at which point rotation has grown sufficient to forbid direct addition of further quanta. Henceforth, evolution may continue only through associations of entire electrons and positrons — a new and comparatively brief round of matter-building which terminates in the construction of the proton. Since, for presumably the same reasons relating to velocity, a proton must also constitute a "maximum" particle, it follows that further evolution will be restricted to the fusion of masses of the general order of protons — a process even more restrictive, and which terminates at the level of the heavier elements deep inside the interiors of massive stars. Complicated somewhat by a constant struggle of proton-neutron complexes to preserve unity, excessive mass may again jeopardize stability by allowing internal motion to become a regulating factor.

As newly constructed electrons and positrons enter the dense and rapidly rotating nuclear region of a condensing primordial cloud, they will be seen to approach from a direction somewhat *perpendicular to the prevailing flow of energy.* Spiraling deeper into the core of the swirling cloud, a discrimination will eventually arise with regard to the direction in which fusion is most likely to be facilitated. Indeed, there will be a distinct preference for recent arrivals to be incorporated at the *leading edges* of developing proton-electron complexes. Not only must an evolving structure encounter more potential members in the direction of its motion, but it will also be more difficult for particles to enter into a relationship from the opposite direction, since the principal body is tending to move away at high speed. (Any antiprotons which are formed would be annihilated through subsequent encounters with overwhelming numbers of protons.)

Now it is obvious that contraction of a primordial cloud, with a steady acceleration in velocity of rotation, cannot continue indefinitely. For while very high density is required at some stage of creation in order to expedite the evolution of material particles, the observed distribution of matter within the universe makes it fully apparent that this must be a rather temporary state — invariably, *a state preceding a mammoth explosion!*

The mechanism triggering this cataclysm must surely involve the tremendous rotary forces which exist deep within the cloud as it reaches an advanced stage of contraction, shortly following a large-scale production of protons. As the innermost regions are accelerated to velocities close to the limiting speed of light, so a strong braking action will arise as material particles begin to resist further acceleration. Intense pressure will build-up, in a manner somewhat analogous to that of a powerful coil spring being wound to maximum tension. (Although a supermassive black hole is readily formed at the center of a contracting primordial cloud, it is prevented from swallowing up all condensing matter forms by reason of extreme rotational velocity which greatly reduces the efficiency of gravitational attraction.) Were it not for the force of material forms still streaming inward, the nucleus of the cloud would be induced to shatter at a relatively early stage. Compressed by the sheer weight of inflowing matter, the region surrounding the core must continue to grow hotter and rotate with ever greater rapidity. For some time this rising outward pressure may thus be contained until, at last, the vast bulk of all matter forms is blasted into space — *from whence the ejected particles will later condense into a supercluster of galaxies!*

Although on a scale far surpassing that of a mere supernova, widely differing circumstances will conspire to mask outward signs of such a stupendous event. The exceedingly powerful gravitational field, combined with innumerable encounters that are highly conducive to fusion due to favorable relative motion, must facilitate the incorporation of all but a negligible fraction of the cloud's total mass-energy into "maximum" particles. With vast numbers of encounters tending to prevent expelled radiation from achieving escape velocity, along with the presence of huge swarms of only partially evolved structures, it will be seen that a very strong incentive exists for free quanta to be promptly absorbed by these undeveloped material forms.

Vestiges of this cosmic matter-building process — implying an unending series of "*Small Bang*" explosions — are surely to be found in the peculiar nature of *quasars*. Since it is most logical to infer that remnants of these superdense sites will persist, in one form or another, it is well to consider related features. The combination of extreme density and substantial mass, which is believed to constitute a unique characteristic of quasars, is in precise agreement with what might be presumed of a nuclear core left over from such a monstrous explosion. Indeed, the presence of a supermassive black hole, at the center of quasar, is exactly what would be anticipated under the prescribed conditions.

What must be considered to be strong support on behalf of such a Small Bang scenario, and crushing evidence against the Big Bang creation model, has surfaced in observation by NASA's orbiting Chandra X-ray Observatory. Revealed in images of the distant galaxy 3C295 (about 5 BLY away), the intense X-ray flux has been ascribed to the presence of a supermassive black hole between two central radiation lobes. It is reported that within a radius of a million light-years from 3C295 there are more than 100 galaxies and enough material to make 1,000 more. Such an object is totally against every form of Big Bang cosmology, since any concentration of gas of this magnitude would surely have been among the very first to have condensed into stars and galaxies! Instead, we have before us matter of incredible mass and density which has failed to condense— even after roughly 10 billion years following a supposed Big Bang explosion! Clearly, astronomers are viewing a recent Small Bang creation site — with a possible antiquity far less than the age of any Big Bang cosmology.

Of all the "mature" or "maximum" particles, so formed within a Small Bang creation site, it is the positron that is most suited to achieve a total escape into regions of future Creation Zones. (It may also be inferred that intense gamma ray radiation — essentially electron-positron associations — may be expected to flow out of such an explosion site, although by the time it completely escapes into the depths of space much of it is likely to be reduced, by reason of myriad collisions, to radiation of the order of X-rays.) Due to a favorable mass-to-velocity ratio, positrons are small enough to be accelerated to escape velocity while yet of sufficient mass to overcome resistence through impacts with surrounding quanta. Protons, on the other hand, cannot be imparted with the same velocity and must be contained by gravitation to form galaxies of stars. Attracted to protons by compatible electric charges, almost all free electrons may be expected to be trapped into association — producing hydrogen atoms, the chief source of stellar fuel. (The production of primordial helium, thought by many scientists to be the product of a Big Bang creation, is now seen to have an explanation in the concept of Small Bangs.)

Not only are we thus enabled to explain the absence of free positrons in our sector of space, but this inferred large-scale ejection of positive charges into regions between the superclusters provides a clue as to how the primordial clouds could always be induced to adopt the specific rotary direction of a proton. In effect, a positron-enriched nucleus will incite elementary forms to partake of this same general motion as they arrive in the more central regions of the swirling cloud. Influenced by the presence of a predominant number of positive charges, the cores of such clouds may be expected to acquire this common rotary trait at a crucial stage of condensation — from whence it is bound to be perpetuated through successive generations of creation.

The formation of neutrons may be looked upon as a process which occurs in the intense heat and density of stellar interiors. When a pair of protons are brought together with sufficient force as to achieve a mutual penetration of halos, the two nuclei will remain in close association within the confines of a reconstituted overall halo with a characteristic proton rotation. By inferring the presence of an electron nucleus (an electron largely stripped of its halo), the stability of such a union may well be explicable in terms of two positive charges exchanging a negative charge. Thus, at any given moment, one of the proton nuclei will have temporary custody of the electron nucleus and may, in a sense, be considered to be the neutron.

The normal tendency of two protons to repel each other will be negated once all threat of violent halo collisions is removed, and when the compatible rotary motion of an electron nucleus is permitted to intercede and serve as a "lubricant" between the surfaces of two proton nuclei that would otherwise clash and generate intense friction. Assuming that only a modest amount of the most minute quanta is retained within an atomic nucleus as localized halos of minor value, energy may be exchanged by the nuclear components in such a way as to produce a favorable supply of useable "neutral quanta." With gravitational/electrical attraction enhanced to an enormous degree, a short-range binding force of considerable strength may be postulated. Yet another feature, contributing to internal stability, is that of the overall halo itself — an energy barrier which must impose the same restraint upon escape as it does with regard to any inward penetration.

How are we to account for this fortuitous appearance of an electron nucleus? In all probability, it is simply *created* at the moment of nuclear fusion — presumably, out of the halo energy enveloping the protons. Indeed, as two such halos are forcibly driven together a certain amount of impacted halo quanta will be compressed between the two spinning bodies and caused to adopt a specific motion — rotary momentum determined by the angle of the intersecting energy flows which, in this instance, will be seen to be opposite that of a proton. When this swirling accumulation of energy reaches a critical mass-density, it will literally result in the creation of an electron nucleus — a "maximum" particle whose evolution terminates with a rejection of further quanta.

Technically, the ability of an electron to neutralize the charge of a proton must embrace the principle of *interference*, whether it be a case of movement through an atmospheric halo or the interaction of nuclei at the center of such an enveloping cloud of energy. In the former instance, the disruptive influence of an oppositely rotating electron halo will act to cancel outward effects of rotary motion. In the case of a neutron, this disturbance of the overall rotary flow is perpetrated from within. By reason of a nuclear electron's rapid orbital motion around a proton nucleus, and in conjunction with inherent tendencies to incite opposite halo rotation, the original halo structure of the proton will be so disrupted that no significant rotary flow of energy will be established. In short, a neutral particle may be defined as either a *cancellation or absence of any specific halo rotation.*

Due to the greatly reduced halos surrounding components of an atomic nucleus, a nuclear electron is probably permitted to spiral inward toward the actual surface of a proton nucleus — from which point the repulsive properties of a "maximum" particle will prevail and, in the process of preventing a direct incorporation of unwarranted energy, will accelerate the smaller body to a higher orbit. In brief, it will be called upon to describe an orbit of some eccentricity as it is periodically repelled after coming into virtual contact with a rapidly spinning proton nucleus. As a result of this highly eccentric movement, which is perhaps reminiscent of the orbit of a periodic comet, the stability of a free neutron must be somewhat in jeopardy, requiring only the impact of an especially large quantum of energy at a strategic moment to literally propel the nuclear electron clear out of the system. As a consequence of inhabiting a compound atomic nucleus, however, a wandering nuclear electron will almost invariably encounter another proton nucleus before it can manage to escape, thereby assuring the stability of the structure.

The unstable family of mesons, displaying masses intermediate between protons and electrons, may well be interpreted as segments of protons — bundles of electron and positron nuclei torn loose following violent collisions involving atomic nuclei. The heavier class of so-called particles known as hyperons, with masses rated in excess of the proton, are evidently larger aggregations of positron/electron nuclei remaining together for the smallest fraction of a second before dispersing from similar impact sites. During such brief intervals (often less than one ten-billionth of a second!) both positive and negative characteristics may be exhibited, along with disturbing motions which would give rise to neutral properties. Tentative associations of this sort, including transient quarks and gluon-type peculiarities, are made possible by the inability of a system to change its mass-energy relationship in zero time.

Even the rather elusive neutrino, a hypothetical particle of no apparent charge and virtually no mass (almost entirely energy of motion), introduced by science in order to balance certain equations relating to nuclear events, must now receive interpretation in terms of ejected gravitational quanta. However, instead of these corpuscles traveling with a variety of extreme speeds, in all probability they are simply specific — and varying — bundles of minute quanta leaving the site of some nuclear transformation at the velocity of light.

Conservation of Mass-Energy

With the phenomenon of gravitation receiving interpretation as the inborn desire of all creation to fulfill its Destiny through the process of *Cosmic Fusion*, we were soon made aware of a serious misconception which had hitherto prevailed with regard to the propagation of radiation. In turn, clarification of this issue served to reveal the Continuous Creation principle to be the foundation upon which a successful cosmology must be based. And yet, as if by way of a paradox, even the explosive origin concept was seen to have merit when modified to embrace a never-ending series of Small Bangs announcing the birth of "maximum" particles. But this unexpected blend of cosmological models raises a number of unanswered questions — not the least of which involves the problem of reconciling the "law of energy conservation" with the idea of *Continuous Creation* and an expanding universe. Another crucial point which demands clarification is that of the *size* of the universe itself. Together, such issues hold the key to a proper understanding of the Great Cosmic Pyramid, and of a truly momentous prediction regarding the fate of our physical universe.

Of itself, the very premise of Continuous Creation involves a contradiction with the firmly established scientific principle of energy conservation — whereby the total mass-energy content of the universe is viewed as an interchangeable but fixed commodity. Inasmuch as a perpetual creation of new matter must imply a steady increase in the size of our physical cosmos, it is clear that stability may only be maintained through *a literal disintegration of an equivalence of existing matter!* Just as there is birth and death in the realm of biological life, so at the level of physical matter there must be a reconciliation of new creation with a precisely opposite factor. Somewhere, and somehow, matter must surely be caused to sever its bonds with the empirical world and vanish just as mysteriously as it made its initial appearance! Invariably, the laws regulating this departure will be seen to be no less revealing than those governing the introduction of infinitesimal "**A**" between the clusters of galaxies.

Nor is there any way of evading the issue of disintegration. Strange though it might first seem, there can be no question of doubt. For instance, if we were to postulate a universe infinite solely in physical dimensions, we would not have a truly infinite universe. Without new creation all stars must, in time, age and die. What, then, will become of our so-called infinite universe? On the other hand, should we attempt to add a creation factor to our universe of infinite dimensions, we are immediately faced with a flat contradiction. How can we possibly place an endless quantity of new matter in a universe that is already infinite in size — without achieving infinite density? Obviously, there can be but one logical solution Only through the premise of a cosmos *finite* in physical dimensions can we hope to escape the dilemma of having no place to put new creation!

In order to reconcile Steady-state philosophy with the idea that the universe is finite in physical size, we must picture creation as a process which transpires — between clusters of galaxies — within the confines of a vast spherical region of space. Outside of this Cosmic Creation Zone no elementary matter forms may be born — *a supposition which very much suggests that matter, as such, is unable to exist in regions so unduly separated from the remainder of creation.* Indeed, since a violation of the "law of energy conservation" may only be avoided by predicting the annihilation of a specific quantity of matter, and in a specific interval of time, a disintegration factor based upon a critical separation or lack of overall communion is an inevitable conclusion. Presumably, the recession of the outermost systems — *at the velocity of light* — forbids their continued association with creation as a whole. If so, *an abrupt disintegration is the fate which awaits those onrushing systems attaining this critical speed limit — essentially, a crucial distance and velocity of recession relative to the center of the universe!*

Deeper insight into this inferred disintegration is likely to be found upon consideration of implications inherent in our revised concept of energy propagation. Invariably, reinstatement of the "law of addition of speeds" must raise the issue as to what really transpires at the precise edge of the universe, where radiation becomes totally trapped by the phenomenon of cosmic expansion.

As light-emitting galaxies of stars — along with such highly concentrated energy sources as supernovae and quasars — reach the cosmic edge, the speed of their inward - bound radiation will be precisely equal to the outward flowing of the universe. As a result, it will become trapped to form an encircling *"shell of energy."* In essence, it will remain essentially stationary — trapped indefinitely in a universe that is infinite in time. The sheer magnitude of this invisible shell is bound to tax the imagination, simply because of the infinity of time aspect, which must allow radiation to add upon radiation. In all probability, *the inferred disintegration is accomplished through violent collision — at the edge of the universe — when material structures (e.g., stars, planets, etc.) strike this enveloping shell of intense energy at the velocity of light relative to the center of the universe!*

So concentrated is this fateful energy field that it may, in fact, be likened unto an *encircling black hole!* Quite invisible to the most powerful telescope, such an impenetrable energy barrier must surely reside at the cosmic periphery, where it acts to balance creation in rather dramatic fashion. Subsequently, it may be deduced that many currently observed star systems exist only in the form of *ghostly images* — the real celestial objects having long since vanished in response to a fundamental law requiring the conservation of mass-energy!

Should we ask for an explanation as to why the "boundary" of the universe happens to be situated where it is, and not in some other location, it can only be concluded that volume is regulated by the *quantity* of spirit comprising one entire generation of "**A.**" In other words, whatever amount of spirit is necessary to constitute the level of Perfection, so the size of the universe must coincide with the need to express all elementary and intermediate levels in terms of a physical manifestation. *Hence, the total amount of matter in existence will be seen to be directly related to the quantity of spirit required to populate all levels of the Cosmic Pyramid.* Accordingly, with knowledge of the actual size of the universe, it will be possible to deduce the full number of"steps" of spiritual doubling (or fusion) which extend from the initial status of a lowly "**A**" to the Ultimate State of Perfection which we would define as God.

What are the chances of our Earth coming to such a dramatic ending in the foreseeable future? Actually, the odds appear to be strongly in favor of just this happening long before biological life is extinguished as a consequence of stellar evolution. Regardless of the size of the cosmos, it may be computed that eight times as much matter resides in the outermost half of the universe (from the point where recession is 50% that of light) as in the remainder — all of which must undergo annihilation in one cycle of Hubble's constant. By the laws of probability, it might be argued, we are presently living on borrowed time and are long overdue!

True though this may turn out to be, the immediate situation is not nearly as frightening as it might first seem. Inside the huge volume of the universe, a human lifetime represents the smallest fragment of time in the overall sequence of cosmic events. Even conceding that we may now occupy a position quite close to the edge of the universe, the prospects of us reaching this critical boundary are extremely slim in terms of a decade or a century. Should we substitute millions of years it would, very conceivably, be entirely another matter.

The Distribution of Quasars

In portraying quasars as ancient creation sites, it may well be prudent to consider the factor of intrinsic luminosity. Conceding the strong possibility that colliding galaxy mergers could produce massive black hole cores and luminosities which mimic quasars, it is perhaps in order to adopt a minimum absolute magnitude of not less than -23.0

as a means of identification. (Since a genuine Small Bang vestige will possess a central black hole of substantially greater mass than a typical large galaxy, it is to be expected that it will be characterized by intrinsically higher luminosity.) Also complicating matters is the realization that quasars tend to be far more luminous while in their youth, when there is much more gas and dust available to fuel their supermassive black hole dynamos. Although its luminosity has dropped enormously, with the passage of time, it is highly likely that the massive galaxy known as M87, at the center of our Local Supercluster, was once a quasar parent of our Milky Way system in the distant past.

Considering that we see quasars as they were billions of years ago, we also see them as appearing in positions distorted by the curvature of their light as it is influenced by expansion of the universe. By way of recapitulation, this displacement of images— due to "cosmic arcing"—is determined by three factors:

(1) The *time* of light transit, as implied by distance. (A longer transit time must result in increased curvature.)

(2) The *position angle* of a quasar relative to our movement toward the edge of the universe. (The greater this angle, the greater the bending of light.)

(3) the *proximity* of an observer to the outer edge, in which the curvature of radiation is seen to be enhanced with increased distance from the center of the universe.

In every instance, displacement will be one *in which light is bent toward the cosmic center*. (See Fig. #3.) The inevitable question which now arises, of course, is whether observation can substantiate such a prediction.

Should we happen to be situated in a relatively central position of the universe (with little motion toward the edge) we might expect to see no appreciable clumping of quasars, which would appear to be evenly distributed in all directions. Upon the far more logical basis of an eccentric location, and assuming the validity of our new cosmological model and revised views of light propagation, one should anticipate a decidedly unhomogeneous pattern with regard to the positioning of quasar images.

Upon detailed analysis of what is generally conceded to be a comprehensive quasar survey, published by the rather prestigious *European Southern Observatory*, [1] a most instructive pattern is noted. Containing a total of 2,720 members (of at least absolute magnitude -23.0, it is at once evident that the vast bulk of quasars reside in two pronounced clumpings. Not only are these two major concentrations situated in exactly opposite directions of the sky, but a highly conspicuous shortage is observed at right angles to such aggregations— posing a distribution which is totally against the dictates of popular Big Bang cosmology! (See Fig. #6 and Fig. #7.) Moreover, when the most luminous of these quasars are considered as a group, it is revealed that an overwhelming proportion of high redshift/high luminosity quasars are centered about the constellation of Sculptor in the Southern Hemisphere— a region which is also seen to be essentially devoid of high luminosity quasars of low redshift! In contrast, the opposite hemisphere is characterized by a mixture of both high redshift and low redshift quasars.

Although it might be thought that the center of the universe must lie in the midst of the high redshift clumping, this is most certainly not the case. Upon the basis of a Small Bang/Steady-state cosmology, and an eccentric cosmic location, the central regions of our universe should display a wide variety of redshifts. On the other hand, the nearby edge is bound to be deficient in brilliant low redshift quasars— simply because there is less room for new and highly luminous quasars to be born! Furthermore, a concentration of high redshift objects (toward the immediate edge) is to be expected by reason of the virtual entrapment of radiation, which allows us to view multiple generations of quasars with greatly extended light transit times. Accordingly, it will be concluded that the cosmic center is probably located near the coordinates of R.A. 13 hr and Dec.+30 degrees — near the border of the constellations of Canes Venatici and Coma Berenices in the Northern Hemisphere. The nearby cosmic edge is likely to reside in the general vicinity of R.A. 1 hr and Dec. -30 degrees— very possibly within the constellation of Sculptor.

Quite inexplicable in terms of any Big Bang interpretation, the mystery of a dichotomous quasar distribution is therefore seen to have a perfectly logical solution. It is clearly much more than a coincidence that the congested central core of the universe is accompanied by a pronounced right angle shortage— invariably, a phenomenon produced by the prescribed "arcing" or bending of extragalactic light. The prodigious edge group clumping, with its preponderance of high redshift quasars, has a ready explanation upon the basis of our eccentric cosmic location and a revised concept of radiation propagation. Instead of visualizing such highly redshifted radiation as having come from the outskirts of the universe, many billions of light-years away, it is really our own Milky Way which has done most of the traveling! Moving outward more than half the radius of the universe since the formation of our galactic system, we are only now encountering the light of long extinct quasars which has literally hung suspended— *almost in a state of total entrapment*— close to the cosmic edge for enormous periods of time. When we observe high redshift quasars, with light transit times of tens of billions of years, we are actually viewing very ancient images which might well be described as *celestial fossils*— the real objects having ceased to exist eons before our Earth was born. Thus it may be presumed that— in the direction of the nearby edge— all quasars (and possibly all galaxies) are merely ghostly apparitions of long extinct phenomena!

With quasar luminosity acknowledged to be a function of age, it is expedient to assess the distribution of quasar luminosity versus redshift over the expanse of the entire sky. In searching out high luminosity quasars of all visual magnitudes, and adopting a lower limit of absolute magnitude -29.0, along with a redshift minimum of Z=2.5, we find a total of 65 qualifying quasars in the aforementioned *European Southern Observatory* catalog. Upon plotting these objects on the two hemisphere charts (Fig. #6 and Fig. #7), it is at once evident that a preponderance (41 to 24) is centered upon the immediate coordinates of the nearby cosmic edge.

Of equal significance is a pronounced shortage of low redshift quasars of high luminosity residing in the vicinity of this same nearby cosmic edge, *where they are quite conspicuous by their absence!* Most abundant elsewhere, they are virtually absent in this one particular direction. Needless to say, an explanation is demanded for yet another strange anomaly. Impossible to answer in terms of any Big Bang cosmology, this peculiar shortage will now be conceded to arise by reason of our highly eccentric cosmic location, as reduced volume of space acts to impose a rather drastic restriction as to when (and in what direction) luminous quasars of recent vintage may be formed. In this one preferred direction of the sky, it must surely follow that the last light of many neighboring quasars has already passed us en route to more central regions of our universe.

Valuable insight into this biased distribution may be obtained through analysis of the quasar sample following its division into a succession of redshift bins. Expressed in multiples of Z=0.200, and considering an edge group member to be one inhabiting the Sculptor quadrant of the heavens, we are enabled to construct the data conveyed in Fig. #8. A truly amazing picture emerges. With virtually zero difference at Z=0.200, substantial excess luminosity (away from the edge) is quickly encountered and observed to peak in the vicinity of Z=0.400—*from whence the noted luminosity differential is seen to proceed in the opposite direction!* Effecting a complete reversal between Z=1.80 and Z=2.00— representing the very point where *time of propagation overtakes distance*—a progressive surplus of intrinsic luminosity (in the direction of the edge) will be seen to characterize the remainder of the higher redshift bins. (See Fig. #8 and Fig. #9.) Nor is this variation of minor consequence. From Z=0.400 to the limit of detection the combined bin differential luminosity translates to a brightness factor of some 360%— a factor which is direction oriented! Most assuredly, this striking pattern is not at all compatible with an explosive Big Bang origin.

In the process of explaining this remarkable observation, it will be seen that we are now in a position to verify a proffered solution for a major weakness inherent in previous versions of Steady-state cosmology— namely, the lack of conservation of mass-energy. Indeed, very graphic confirmation of this momentous prediction (involving annihilation at the cosmic periphery) is already at hand in the evidence afforded by Fig. #8, in which there is clearly a marked tendency of intrinsic luminosities and extreme redshifts to be noticeably higher in the direction of the nearby edge. Inasmuch as the energy output of a quasar is bound to diminish with age (typically, a hundredfold reduction may be inferred in the span of about 10 billion years), as it slowly evolves into a giant galaxy with a supermassive black hole core, it follows that *higher luminosity is also a definite indication of youth!* But why should they be younger? This is surely the question which begs an answer. The obvious conclusion is that something must be preventing normal aging of quasars in this one region of the sky. Invariably, *it is a selective effect imposed by premature disintegration at the boundary of our physical universe*— from whence an equivalence of mass-energy is subsequently reincarnated as myriad infinitesimal quanta in the great voids of space.

D/R Arcing of Radiation

The very premise of distance/repulsion (D/R) arcing of radiation affords a variety of observational tests. In addition to the large-scale phenomenon of "cosmic arcing," in which distant quasars are revealed to have a most peculiar dichotomous distribution, we are led to infer yet another and more localized instance which might be termed "galactic arcing."

Impressive statistics have been advanced which clearly indicate that a disproportionately large number of quasars are to be found in close angular proximity to relatively nearby massive galaxies. Quite inexplicable upon the basis of traditional views, such an association suggests a physical connection, which would not be compatible at all with the principle of redshifts serving as reliable distance indicators. It therefore remains to be explained how this anomaly can be resolved in the face of strong conflicting evidence that at least some quasars are at distances which are essentially commensurate with their redshifts.

Notwithstanding that the proposed cosmic aberration factor is able to account for the opposed clumping of quasar images, it is evident that this general cosmic arcing will be insufficient to explain such a quasar/galaxy anomaly. Yet another subtle mechanism must somehow be involved— *one which would permit these seemingly associated galaxies to focus incoming quasar radiation toward such centers of mass!* In all probability, it is the proffered D/R factor which is responsible for creating a scenario of high redshift quasars appearing to be connected with low redshift galaxies.

Just as a steady outward flowing of D/R quanta must produce an expanding universe and a large-scale displacement of images toward the cosmic center, so a massive intervening galaxy is able to induce small-scale curvature in the path of radiation approaching from a remote source— in this instance, curvature away from the less distant concentration of mass. However, the very action of "repulsive" quanta pushing against such approaching radiation must ensure that whatever does reach us will do so from a somewhat different angle— namely, *from a direction that is considerably closer to the intervening galaxy!* (See Fig. #10.) Subsequently, what appears to be an overabundance of quasars— in angular proximity to major galaxies— is really an illusion produced by D/R arcing, and this hitherto unrecognized aspect of gravitation is thus seen responsible for another example of celestial deception.

Strong statistical evidence exists in support of the concept of "galactic arcing," by which the aforementioned D/R factor is seen to be responsible for producing the illusion of a disproportionate number of quasars seemingly associated with massive galaxies. In order to achieve this clumping, in the angular distribution of quasars, it is necessary for these images to have been displaced from regions which are adjacent to an intervening galactic body. Remarkable confirmation of this prediction may be found in the research of several prominent astronomers, [2] in which a dearth of quasars is indeed shown to characterize the outlying regions surrounding just those massive galactic systems in question.

Among the many astronomical anomalies to be uncovered are

instances where luminous bridges appear to physically connect two objects of widely differing redshift— thereby negating the worth of extragalactic redshifts in formulating any distance/velocity scale. There is now every reason to believe that a solution to such enigmas may be found in the concept of *cosmic arcing*, as imposed by the D/R factor responsible for our expanding universe.

One especially notable example of a luminous bridge is a seeming link between a galaxy known as NGC 4319 (Z=1,700 km/s) and the quasar Markarian 205 (Z=21,000 km/s). Located at the coordinates of R.A. 12 hr-21' and Dec. +75-45' degrees, the quasar is situated almost due south of the galaxy. Close examination of this pair reveals a narrow but quite distinct connecting bridge of luminosity. At first glance, this would appear to afford irrefutable proof of physical proximity, being located at the edge of the halo of gas and stars surrounding the galaxy's nucleus. It is highly probable that this illuminated bridge is simply a product of cosmic arcing. In the course of having its radiation bent toward the center of the universe, the distant quasar's light will be seen to have passed behind the nucleus before emerging in the vicinity of the galactic halo. During this grazing passage any intervening matter will be light-enhanced, in the form of a narrow illuminated beam or path, by the relatively intense radiation of the quasar. It may well be much more than mere coincidence that this bridge points precisely toward the cosmic center!

Incidentally, this general deflection of images— toward the center of the universe— may explain a curious distribution of quasars about the galaxy M82. (The coordinates of this system are R.A. 9 hr-51' and Dec. +69-55' degrees.) No less than four quasars appear to lie in a somewhat narrow cone stretching to the southeast of the galaxy— a direction which just happens to coincide exactly with a line extending toward the cosmic center! It is difficult to refute the inference that we are looking at an example of cosmic arcing reinforced by a more localized galactic focusing of images. Needless to say, such instances serve to emphasize the importance of conducting an unbiased search for other celestial examples confirming this most logical interpretation.

Nor is this general cosmic arcing of images confined to distant quasars. On the contrary, should our Milky Way be situated in close proximity to the cosmic edge— as all evidence strongly suggests— we would then be in a position to observe a most peculiar clumping of galaxies. In effect, the outflowing stream of D/R quanta will cause the images of many right- angle/edge-group exterior systems to be shifted considerably toward the center of the universe— producing the mirage of elongated sheets of galaxies stretching far across the heavens. It is to be noted that such a phenomenon has already been discovered by astronomers, who refer to this bizarre distribution as the "Great Wall." (Incidentally, the so-called "Great Attractor" is likely a companion effect of cosmic arcing. This rather curious galactic streaming has led some astronomers to postulate a mysterious unseen attractive force lurking somewhere in the depths of space. However, it will now be seen that a more appropriate title might well be the "Great Illusion.")

References:

[1] *A Catalogue of Quasars and Active Nuclei (2nd Edition)*, Scientific Report #4, April 1985, M.P. Veron-Cetty and P. Veron

Note:

While a more recent and greatly expanded quasar catalog has become available, containing some 23,760 quasars brighter than magnitude -23.00, a detailed analysis of this *Veron + 3001 Quasar Catalog* will be seen to fully substantiate conclusions drawn from studies of all previous catalogs. Indeed, in this new catalog the pronounced dichotomous clumping of quasar images is seen to be virtually double that of the older listing! (Any reduction noted in average luminosity of edge-group quasars, in this new catalog, is readily explained by the fact that there was biased concentration on this one very fruitful location— resulting in detection of disproportionately larger numbers of faint quasars.)

[2] Chief among astute scientists to pioneer such research is Halton C. Arp, whose works have inspired considerable controversy

and who is certainly deserving of much greater recognition. Back in the 1980s, this courageous astronomer risked peer censure and career by championing the issue of discordant redshifts— a highly relevant phenomenon which will be seen to supply the key needed to unravel many of nature's more puzzling mysteries.

5

The Nearby Cosmic Edge

Redshifts as Distance Indicators — Discordant Redshifts — Our Proximity to the Edge— "Quantization" of Redshifts — The Enigma of Galactic Redshifts — The Microwave Background

To better determine our present position with regard to what now looms as a rather fateful cosmic edge to the universe, it is necessary to evaluate certain data which has become available through astronomical observation. In particular, the issue of celestial redshifts, along with that of the microwave background, must receive proper interpretation.

Redshifts as Distance Indicators

In order to assess the true worth of extragalactic redshifts, as a viable means of determining distance, it is helpful to review certain fundamentals that are implicit in our revised concept of radiation propagation. First and foremost, a redshift affords an accurate measurement of the *elapsed time* between emission and eventual detection — quite regardless of direction and any conceivable form of intervention experienced by radiation while it is en route. In order to convert time into distance it is essential that the factor of cosmic expansion be taken into consideration, as radiation is caused to propagate at the velocity of light *relative to the moving field in which it is embedded!* With both direction and eccentricity of our cosmic location regulating the approach of radiation, we are led to contemplate such issues as luminosity and angular displacement in terms of redshift.

The luminosity of an object may be described as a measure of the number of photons that are received in unit time. Assuming no intervention, the traditional inverse-square law will apply to the actual distance between source and observer's "location" at the time of light emission. In this instance "location" is defined as that region of space which, in full compliance with the dictates of cosmic expansion, would intercept such radiation in the interval of time reflected by redshift — regardless of whether this region is occupied by receiver's world at the moment of emission. Restated slightly, with "location" situated in the direction of the cosmic center, distance will depict the time required for expansion to move it to the position of a source residing toward the edge. With regard to a source located at the other side of the universe, distance will express the propagation of radiation through changing flows of quanta. (After passing through the cosmic center a "headwind" will be replaced by a "tailwind.")

The angular displacement of an object is analogous to luminosity, also being commensurate with the inverse-square law applied over the distance between source and receiver's "location" — within a steadily expanding universe — at time of light emission. Accordingly, angular size will not change with an observer's proximity to the cosmic edge, and will therefore always exhibit the same relationship to distance as luminosity.

Extragalactic redshifts, however, while affording an accurate picture of elapsed time of transit, are seen to be highly susceptible to misinterpretation when one attempts to convert measurements to velocity of recession and distance — leading to numerous cases of confusion and outright contradiction with regard to a variety of celestial associations.

Compounding any redshift/distance relationship is yet another hitherto unrecognized ingredient. Contrary to abundant evidence, and ignoring the research of certain astute scientists, there has been a general reluctance to accept the idea that interstellar hydrogen gas could act to enhance redshifts. [1] The fact that radiation is known to propagate faster in a vacuum than through other mediums, such as air or water, is conveniently glossed over by a scientific community quite unwilling to jeopardize traditional Big Bang redshift/recession statistics. In the last analysis, the only issue at stake is really one of assessing the magnitude of this additional "time delay" factor.

Although it might be tempting to theorize that the redshifts of all distant galaxies and quasars could be explained on grounds of their radiation interacting with extragalactic hydrogen, there is good reason to believe that any enhancement must be rather limited — if not in specific and localized instance, then at least in overall extent. Quite simply, the distribution of such hydrogen gas is bound to be far from homogeneous. Indeed, upon the basis of the "New Cosmology," it is to be inferred that the vast bulk of this gas will be confined to relatively small pockets — largely in the vicinity of recent Small Bang creation sites. Thus, it may be deduced that a modest degree of such radiation delay undoubtedly exists, where it must add to the redshifts of distant celestial phenomena. It is likely, however, that instances of extreme redshift enhancement are confined to infrequent occasions where radiation has passed through exceptionally dense concentrations of extragalactic hydrogen gas. (Conceivably, this could afford a logical explanation for certain otherwise enigmatic astronomical anomalies.)

Discordant Redshifts

In a classic instance of serendipity, the very existence of certain anomalous galactic redshifts presents a much needed opportunity to more accurately determine our proximity to the nearby edge of the universe. In effect, a relatively small radial separation — between gravitationally entwined systems residing close to the cosmic edge — is capable of displaying highly discordant redshifts. Hence, calculation of true separation, as a function of respective light transit times, offers a means of deducing our present location with respect to a fateful cosmic boundary. It therefore becomes prudent to examine as many of these interacting systems as possible, in expectation that a truly plausible agreement as to distance will be forthcoming. [2]

Although a great preponderance of discordant redshifts are to be found in the direction of the nearby edge, there are a number of anomalies — toward the cosmic center — which invite discussion. Perhaps the most impressive example discovered to date is that of the remarkable chain of galaxies known as VV 172 (R.A. 11hr-29', +71°-6' Dec.). While four of the members possess redshifts ranging from 15,673 to 16,252 km/s, the fifth system exhibits a rather discordant 37,062 km/s. Since visual observation would lead one to assume that they are all members of the same cluster, an explanation is clearly desired.

In this instance, we are probably dealing with only a modest degree of radial distortion, but considerable angular displacement toward the center of the universe. It is quite conceivable that the discordant member of the system has had its light bent in such a way as to cause it to appear as a fifth component in this unusual string of galaxies. In fact, it is to be expected that the phenomenon of a D/R force must produce many seemingly bizarre associations, including the publicized curiosities known as Seyfert's Sextet and Stephan's Quintet — configurations defying a traditional interpretation, but which have a logical explanation in terms of the "New Cosmology." (Seyfert's Sextet, in which one galaxy member is revealed to possess a redshift almost 16,000 km/s higher than the others, may be viewed as a classic example of cosmic arcing so displacing an image as to create a spurious interloper.)

Stephan's Quintet, however, poses an intriguing example of anomalous redshifts. Located at somewhat less than a right angle from the cosmic edge (R.A. 22hr-34', Dec. +34°-0'), this group should really be looked upon as edge inhabitants which have been subjected to considerable arcing. Three of the systems possess redshifts of 6,700 km/s. A fourth has been measured at 5,700 km/s; while the remaining galaxy exhibits a redshift of only 800 km/s — seemingly in complete disagreement with the others and contrary to all signs of gravitational interaction. Also adding to the overall scenario is the presence (some 20' of arc away) of a large neighboring spiral galaxy known as NGC 7331, with a redshift similar to that of the low redshift member of the quintet. In turn, this major system would seem to have three small companions of redshifts 6,300, 6,400 and 6,900 km/s. When it is further considered that there exists a diffuse bridge of radio-emitting material between NGC 7331 and Stephan's Quintet, the entire picture begins to resemble a celestial zoo.

Certain aspects of this Stephan's Quintet mystery, which has persisted for several decades, are now capable of a measure of clarification as a result of recent observation by the Hubble Space Telescope. [3] The very fact that this instrument was able to resolve individual stars, in only the 800 km/s galaxy, is a strong indication that this galaxy is indeed much closer to us than the high redshift members. There is, of course, two possible explanations. First, NGC 7320 is simply a foreground object that just chances to lie in close angular displacement with respect to the more distant systems. The alternative is to concede that cosmic arcing has served to bend the light of the more distant galaxies to the extent of causing their images to appear as companions to relatively nearby NGC 7320.

With the advent of a new generation of more powerful telescopes, it should be possible to find irrefutable evidence of many highly discordant galactic redshifts in the direction of the cosmic edge — evidence which promises to support many fundamental premises inherent in the "New Cosmology." For example, should just one instance be discovered of individual stars being resolved to the same degree — among gravitationally connected systems exhibiting discordant redshifts — it will confirm the inferred "compression of images" due to increased time of radiation propagation, as such corpuscles fight against a "headwind" of outward flowing quanta.

(Hopefully, this challenge will not be ignored by the scientific community, and astronomers will soon find themselves motivated to undertake a systematic search of the heavens in the direction of the nearby cosmic edge.)

Our Proximity to the Edge

Foremost in our minds must be the question of Earth's present location in space, as it rushes toward oblivion at the periphery of the universe. The task confronting us is to assess the actual degree of retardation (or time delay) imparted to radiation reaching us from remote sources situated in the direction of the nearby cosmic edge. If this key factor can be determined, with reasonable certainty, it should be possible to calculate our proximity to this impenetrable boundary as a function of such distortion.

Two basic approaches to this problem appear feasible. One method consists of examining a sample of highly elliptical galaxies (in the direction of the edge), so oriented that one end is pointed toward this fateful boundary. By reason of the near entrapment of radiation— at the cosmic edge— the side closest to the edge will invariably display an artificially enhanced redshift. In turn, this must result in an apparent increase of galactic rotation— hence "excessive" mass— for the system. If one is able to properly estimate the true radial (line of sight) distance traversed by radiation, from opposite ends of such samples, calculation should reveal the degree of time retardation characterizing the very region of space that we now occupy. Once a determination is made of this key ingredient, simple analysis of pertinent celestial phenomena will afford an answer to the momentous question as to how much time remains before our planet's rendezvous with destiny.

A second method of determining this vital retardation factor involves a study of cosmic edge galaxy associations featuring discordant redshifts. By making a more realistic assessment of the true radial separations, among components of such gravitationally connected systems, it becomes feasible to deduce the amount of time delay imposed upon radiation as it attempts to propagate against the outward flow of D/R quanta. Although a measure of uncertainty is bound to be inherent in estimating genuine radial separations, it will at least afford a "ballpark" idea of the distortion factor involved. Armed with this crucial factor, one may likewise determine our proximity to the cosmic edge.

While acknowledging that a study of many discordant redshift associations will be required in order to reduce error, evidence does exist which will allow us to proceed upon both fronts.

NGC 450 and UGC 807:

By a stroke of irony two galaxies, appearing visually to be in close contact but long considered to be widely separated in radial distance, may well provide the means to deduce this key time retardation factor. Situated in the immediate direction of the cosmic edge, NGC 450 appears to be a typical spiral galaxy viewed almost face-on and associated with a much smaller elliptical system known as UGC 807. The smaller galaxy is inclined to us at an angle, with one end of this elongated companion pointing in the general direction of the nearby edge. What has served to convince mainstream astronomers that this optical pair is widely separated in space is their very different redshifts. While NGC 450 has a redshift of some 1,863 km/s, UGC 807 is clocked at 11,587 km/s. Traditional interpretation of recession velocity into distance would translate into some 111,780,000 LY and 695,220,000 LY, respectively, based upon a universe with an 18 BLY radius. Thus a separation of about 583,440,000 LY is generally believed to exist, and to rule out all possibility of physical interaction. (See photo.)

Yet another strong argument, implying a separation measured in hundreds of millions of light-years, is that of rotational velocity. In compliance with a firmly established law of physics, galaxies of high mass must invariably rotate at higher speeds. The rotation rate of NGC 450 is of the order of 145 km/s; while the observed figure for UGC 807 is some hundreds of km/s greater— indicative of a much larger and more massive galaxy! Seemingly, therein lies further crushing evidence against the pair residing anywhere close to each other. In short, the rotational velocity of UGC 807 implies a massive system which would have to be hundreds of millions of light-years more remote in order to appear visually only one-third the size of NGC 450. Moreover, should we place the pair in close association, with mass/size/luminosity commensurate with rotational velocity, UGC 807 would be so big that it could literally swallow NGC 450. Subsequently, in the eyes of the vast majority of astronomers, there was no question but what the two galaxies must be widely separated in space.

However, with knowledge of how radiation is really propagated through space, along with insight into a more realistic cosmological model, it is now prudent to revisit this intriguing pair of galaxies. Once it is realized that an enormous distortion of redshifts exists in the vicinity of the cosmic edge, by reason of the near entrapment of radiation, the stage is set for a major discovery. Indeed, it becomes feasible to refute the two principal arguments against close association of NGC 450 and UGC 807— while also allowing us to make a reasonably accurate determination of the key retardation factor prevailing in this region of space.

Crucial to a resolution of this factor is an estimation of the true radial separation of this optical pair. The UGC 807 system must now be recognized as a rather small elliptical galaxy, possibly no more than 10,000 LY across, and with a radial separation from NGC 450 that could be no more than a few thousand light-years. In assessing this issue of excessive rotational velocity, it may be computed that a galaxy with the corrected size and mass of UGC 807 should exhibit a rotation speed of only 75 km/s, instead of the measured hundreds of km/s that astronomers would use to infer a much higher mass/luminosity/size figure. Tending to complicate matters somewhat is the direction of rotation of UGC 807, which could be either toward or away from our line of sight. If toward us, then to determine the factor of enhanced redshift we would have to add 75 km/s to any observed figure. If intrinsic rotation is away from us, this 75 km/s must be subtracted to give the system's true rotation.

Proceeding upon such a basis, it is informative to examine statistics relating to the noted excess rotational velocity of UGC 807, a study of which was published in *The Astrophysical Journal*, 271 (pages 556-563), in an attempt to disprove Arp's contention of discordant redshifts. In two plates (#1808 and #1813), multiple redshifts were obtained over the expanse of the elongated galaxy. In one plate there are 27 measurements, which range from 11,297 km/s (NE corner) to 11,705 km/s (SW corner). The other plate contains 24 measurements, ranging from 11,268 km/s (NE corner) to 11,689 km/s (SW corner). The average redshift differential translates to some 415 km/s—increasing steadily from NE to SW. What is so significant about this study is that it is the high-redshift SW end of the galaxy which points almost directly toward the nearby cosmic edge— precisely what must be inferred from principles inherent in the "New Cosmology," which requires radiation closer to the edge to be the most affected. (Had this increase proceeded in the opposite direction, it would have posed a serious contradiction to theory!)

In pursuing an attempt to deduce a radiation delay factor, from a revised distance/time estimate of the light reaching us from opposite edges of UGC 807, it is perhaps in order to consider the excess redshift differential (either 490 km/s or 340 km/s) in terms of the velocity of light (300,000 km/s), and in conjunction with the inferred radius of the universe (18 BLY). Thus such a delay factor would receive expression as the excess redshift differential divided into the velocity of light and multiplied by the radius of the universe, which is then divided by an estimated radial light transit distance.

Expressed in terms of formula:

$$\text{Time delay factor} = \frac{ER}{VS}$$

Where:

E = Excess Redshift Differential (km/s)
V = Velocity of Light (km/s)
R = Radius of Universe (LY)

S = Radial Separation (LY)

By way of example, should we tentatively postulate a radial light transit distances of 100 LY and 150 LY, we may derive the following statistics (rounded figures) for this crucial time delay factor:

100 LY Radial Separation:

490 km/s (+ rotation) = 300,000 delay factor
340 km/s (- rotation) = 200,000 delay factor

150 LY Radial Separation:

490 km/s (+ rotation) = 220,000 delay factor
340 km/s (- rotation) = 153,000 delay factor

Our present location, relative to the cosmic periphery, may be found by dividing the radius of the universe by this same radiation time delay factor. (Note: a reduction in this time delay factor, by increasing the radial light transit distance across UGC 807, must translate into increased distance from the cosmic edge.)

Should we be so bold as to adopt a retardation factor of the order of 300,000, this would tend to imply that our immediate region of space is likely situated some 60,000 LY from the black hole barrier so believed to enclose the universe. Alternatively, should we substitute a delay factor of 200,000, it would be indicative of a current location some 90,000 LY from the nearby impenetrable cosmic boundary. Hopefully, a more precise figure will be forthcoming with additional study.

Armed with what might well be considered a rather simplified but plausible estimate of the magnitude of the near entrapment of radiation, in the region of space under consideration, we may now contemplate the true radial separation between NGC 450 and UGC 807. Inasmuch as it will be seen to have taken 583 million years for this radiation to traverse a distance that has been reduced enormously by a deduced radiation delay factor, it is expedient to put such a factor to test. Upon the basis of a retardation factor of some 300,000, this would imply a true radial separation between these two galaxies of about 2,000 LY. By the same reasoning, a delay factor of 200,000 would give a figure of roughly 2,900 LY, etc. (While a separation of only two or three thousand light-years may seem a trifle close, in view of the degree of observed physical interaction, it must be understood that gravitational influence is the product of both proximity and *time* of proximity, as well as mass. In this instance, there is every indication that the two systems have only recently approached each other.)

The immediate task now before us is to see if this deduced time delay factor is able to provide realistic separations with regard to other physically connected galactic systems— specifically, those which exhibit discordant redshifts and reside in the vicinity of the nearby cosmic edge.

The NGC 7603 System

In pursuing a study of such examples, one especially promising association involves NGC 7603, a large Seyfert galaxy with a smaller companion system evidently attached by a curving filament of luminous stars and gas. The redshift of the larger galaxy is indicative of a traditional recession velocity of some 8,700 km/s (Z=.029); while that of the smaller galaxy is 17,000 km/s (Z=.067). Upon the basis of traditional views of radiation propagation, a radial separation of some 504,000,000 light-years is indicated— clearly an absurd distance for two systems displaying every sign of recent contact! Application of a radiation retardation factor of 300,000 would require a radial separation between NGC 7603 and its companion of roughly 1,700 LY in order to equate theory with observation. In any event, such an adjusted separation is certainly far more plausible than what has long been accepted by mainstream astronomers. (See photo.)

Rather startling observational evidence, concerning the NGC 7603 system, has come to light via the research of Halton C. Arp, a leading astronomer and cosmologist. His observations afford a new and somewhat serendipitous opportunity to deduce our present location with respect to an inferred fateful edge of the universe. It also offers a way to put the theorized radiation retardation factor to a crucial test.

Located within a relatively narrow band of stars and gas, extending from NGC 7603 (Z=.029) to its smaller companion (Z=.057), lie two much higher redshifted spherical objects of Z=.243 (Object #2) and Z=.391 (Object #3). Although it is tempting to assume that these two recently detected objects are simply distant quasars, whose images have been displaced by a combination of "cosmic arcing" and "galactic arcing," there are very convincing arguments to the contrary.

A most rational assumption for the luminous bridge of stars and gas, which connects the two galactic systems, is to presume a grazing passage (some millions of years ago) of the smaller galaxy with NGC 7603. During this encounter a considerable quantity of stars and gas was drawn from the extremities of the large Seyfert galaxy. What needs to be explained, upon the basis of any quasar interpretation of the two high redshift objects, is just how they both managed to lie within this relatively narrow band. Even allowing for "arcing" of their radiation, it is quite against the odds that both images should appear embedded in the midst of such a narrow connecting stream of stars and gas.

One promising alternative to a quasar interpretation of Object #2 and Object #3 is to think in terms of globular clusters. Inasmuch as such spherical clusters of stars are to be found in great numbers orbiting major galaxies, it is reasonable to assume that— in the process of a grazing passage— one or more of these clusters would be caught up in the companion's gravitational grip and caused to pursue a trajectory consistent with the plane of interaction. Furthermore, it is logical to presume enhanced luminosity on the part of any globular cluster so affected, either by reason of mergers with other clusters or through incorporation of nearby stars and absorption of interstellar gas.

In this event, a viable explanation would seem to exist for the observed redshifts of Object #2 and Object #3. The fact that #3 is the more redshifted one is not nearly as peculiar as might be thought, since it would simply require a higher acceleration toward the direction of the nearby cosmic edge. With the passage of time, a relatively small initial difference of velocity and distance will lead to a large magnification of redshift and time of light transit. (The closer we are to the cosmic edge, the more intense this retardation will be—

literally reaching a factor of millions, or even billions, for an observer situated at the very periphery of the universe!)

On the whole, it may well be concluded that the globular cluster scenario is likely to prove the most credible solution for the two anomalous objects which have been detected within this narrow band of luminosity extending from NGC 7603 to its smaller companion galaxy. What now requires urgent investigation is how the trajectory of this connecting band— which shows surprisingly little variation of redshift over most of its length— can be utilized to make a reasonably accurate determination of our proximity to the cosmic edge.

The very fact that there is found to be so little deviation of redshift, along the band of connecting luminosity, is a strong indication that the grazing event with NGC 7603 must have occurred at a rather flat angle to our line of sight. As we shall see, this deduction does not bode well for the long term prospectus of our region of space— including that of Earth! It is more than a little disturbing to note that such comparatively low implied radial separations are characterized by so high a redshift differential. (To the extent that a smaller radial separation is indicated, so closer proximity to the cosmic edge must be inferred.)

Upon assuming a radiation delay factor of either 300,000 or 200,000, the following statistics (rounded figures) may be derived relating to true radial separations of the four anomalous objects in question, of which there are six possible relationships:

NGC 7603 (Z=.029) and Companion (Z=.057)

Radial separation is Z=.028, or 504,000,000 LY (uncorrected distance). If the retardation factor is 300,000, then their true radial separation is: 504,000,000 LY divided by 300,000 = 1,680 LY. If this factor is 200,000, then such separation would be 2,520 LY.

Object #2 (Z=.243) and Object #3 (Z=.391)

Radial separation is Z=.148, or 2,664,000,000 LY (uncorrected distance). If the retardation factor is 300,000, then their true separation is: 2,664,000,000 LY divided by 300,000 = 8,880 LY. If this factor is 200,000, then such separation would be 13,320 LY.

Companion (Z=.057) and Object #2 (Z=.243)

Radial separation is Z=.186, or 3,348,000,000 LY (uncorrected distance). If the retardation factor is 300,000, then their true separation is: 3,348,000,000 LY divided by 300,000 = 11,160 LY. If this factor is 200,000, then such separation would be 16,740 LY.

NGC 7603 (Z=.029) and Object #2 (Z=.243)

Radial separation is Z=.214, or 3,852,000,000 LY (uncorrected distance). If the retardation factor is 300,000, then their true separation is: 3,852,000,000 LY divided by 300,000 = 12,840 LY. If this factor is 200,000, then such separation would be 19,260 LY.

Companion (Z=.057) and Object #3 (Z=.391)

Radial separation is Z=.334, or 6,012,000,000 LY (uncorrected distance). If the retardation factor is 300,000, then their true separation is: 6,012,000,000 LY divided by 300,000 = 20,040 LY If this factor is 200,000, then such separation would be 30,060 LY.

NGC 7603 (Z=.029) and Object #3 (Z=.391)

Radial separation is Z=.362, or 6,516,000,000 LY (uncorrected distance). If the retardation factor is 300,000, then their true separation is: 6,516,000,000 LY divided by 300,000 = 21,720 LY. If this factor is 200,000, then such separation would be 32,580 LY.

In rounded figures, some plausible radial separation statistics (in light-years) for the NGC 7603 system, follows:

Systems	Factor 200,000	Factor 300,000	Factor 400,000
NGC 7603 and Companion	2,520	1,680	1,260
Object #2 and Object #3	13,320	8,880	6,660
Companion and Object #2	16,740	11,160	8,370
NGC 7603 and Object #2	19,260	12,840	9,630

Companion and Object #3	30,060	20,040	15,030
NGC 7603 and Object #3	32,580	21,720	16,290

Thus it would appear that the evidence now points to a rather truncated future for mankind— perhaps, focusing upon a figure of some 60,000 or 90,000 years? If a retardation factor of 400,000 or more is substituted, it would imply that we have only 45,000 years (or less) remaining. It is to be expected that study of a much larger sample of associated galactic systems— displaying discordant redshifts and residing in the direction of the nearby edge— will reduce present uncertainty. (Indeed, once a sufficient number of such connected systems have been subjected to analysis, any complication posed by our inability to determine intrinsic direction of rotation may be ignored through the process of averaging.)

Other Cosmic Edge Examples:

Due chiefly to the courageous efforts of astronomer Halton C. Arp, whose work has generally been dismissed by short-sighted members of the scientific establishment, a wealth of information has been made available concerning discordant redshifts. Upon the basis of theory inherent in the "New Cosmology," it is perhaps in order to list a few additional examples of distorted redshifts of galactic associations residing in the vicinity of the cosmic edge.

NGC 1232

Located at R.A. 3 hr. and Dec. -20 degrees is a large spiral galaxy (redshift 1,776 km/s) known as NGC 1232. At the end of one of the spiral arms lies a small companion system with a redshift of 6,552 km/s. (See photo.) Although there is evidence that they are physically connected, mainstream astronomers have persisted in their view that they must be separated in space by distance fully commensurate with traditional redshift interpretation— namely, by a radial separation of some 287,000,000 LY. We can now see how the radiation retardation factor has served to drastically distort redshifts. In this instance, a factor of 300,000 would translate into a line of sight separation of about 960 LY; while a factor of 200,000 gives 1,430 LY— clearly a far more realistic scenario than what would otherwise be the case. (Incidentally, a logical explanation is forthcoming for a second apparent companion to NGC 1232. Displaying a redshift of some 26,000 km/s, this compact object would seem to shine through the disk of the nearby galaxy, which should serve to redden it considerably. To the surprise of many astronomers, it is observed to be quite blue in color! There is now every reason to suspect that it is indeed a distant background system which, due to myriad encounters with D/R quanta, has had its light bent in such a manner as to project an image in front of the galaxy's spiral arm. Most assuredly, this proffered explanation is able to resolve yet another discordant redshift mystery.)

AM 0328-222

Situated in the region of the nearby cosmic edge, AM 0328-222 is a spiral galaxy with what appears to be an interacting companion located close to one of its spiral arms. The very fact that there is a redshift differential of some 17,925 km/s between them has caused mainstream astronomers to dismiss the possibility of physical connection, since traditional interpretation would imply a radial separation of about 1,075,500,000 LY. Applying a radiation retardation factor of 300,000, we get a true separation of roughly 3,580 LY; while a factor of 200,000 gives a figure of some 5,380 LY, etc. It may well be concluded that this is yet another instance in which galactic redshifts have been dramatically distorted in the vicinity of the cosmic edge. (See photo.)

AM 2006-295

Lying in the immediate region of the nearby cosmic edge, AM 2006-295 is an unusual two-arm open spiral system with what appears to be a slightly curved elongated galaxy situated within one arm of the main galaxy. (See photo.) On the whole, it presents the appearance of a three-armed spiral (or two merging systems) which should not exhibit a discordant redshift. However, to the astonishment of observers, there is a difference of 22,350 km/s. Given traditional interpretation, this would actually be indicative of a staggering 1,341,000,000 LY radial separation. Application of a radiation retardation factor of 300,000 would imply a true separation of about 4,470 LY; while a factor of 200,000 would give a radial separation of 6,700 LY. (Were we to assume a real separation of 1,341,000,000 LY, it would make the interloping third arm of a size and luminosity that would most certainly border upon the absurd!)

AM 2054-221

The system known as AM 2054-221 appears as a large closed-ring spiral which has just been grazed by a smaller companion. In fact, it gives every impression of a more massive galaxy tearing loose a band of stars and gas from the trailing edge of the smaller system as it passed by. (See photo.) But instead of displaying comparable redshifts, there is observed to be a differential of some 36,460 km/s. A recession velocity of this magnitude would normally be translated into a radial separation of a whopping 2,173,400,000 LY. However, upon the premise of a radiation retardation factor of 300,000, this enormous distance is reduced to approximately 7,240 LY; while a factor of 200,000 gives a true separation of some 10,870 LY. Once again, a most necessary adjustment of all distance/velocity/redshift statistics, in the vicinity of the cosmic edge, is clearly indicated and is capable of resolving many celestial mysteries.

In view of this cosmic edge "compression of images," it may well be asked why there are not more reported instances of discordant redshifts among cluster members. Quite simply, the answer would seem to be that such instances do exist, but they are generally unnoticed because many redshift differences are so great that they appear to require membership in other galaxy clusters! It is only when close angular proximity is noted that the issue of anomalous redshifts is raised. Also a strong contributing factor is the reluctance of many astronomers to seek evidence which would challenge the credibility of long accepted views involving the cherished redshift/distance scale.

(Should posterity come to accept the momentous prediction of impending extinction at the cosmic edge, it will be conceded that it was only made possible through the astute observations and courage of Halton C. Arp. In spite of rejection by many members of the scientific community, his pioneering research into the phenomenon of discordant redshifts must be recognized as one of the greatest achievements in the annals of astronomy.)

"Quantization" of Redshifts

Adding to the mystery of galactic redshifts in general is data which would seem to establish a rather peculiar pattern with regard to cluster members. To the astonishment of many astronomers, it has been disclosed that the redshifts of most galactic systems tend to congregate about certain discrete values, implying that an unknown subtlety is involved in the creation of such redshifts.[4] Understandably, this revelation has served to add impetus to the controversy regarding the usage of redshifts as reliable distance indicators.

Although much remains to be studied in depth, it now appears that this phenomenon is rooted in one aspect of D/R arcing. In the process of interacting with radiation, the D/R quanta must imprint a *"time delay"* which is fully commensurate with this intervention and quite independent of any arcing which may occur. In effect, when radiation approaches our Milky Way system it encounters a specific strength — related to the mass of our Galaxy — of the "repulsive" quanta streaming from our system, which acts to slow it down and to *increase the time of transit.* (The greater the mass of a receiving galaxy the greater must be this retardation factor that is imposed upon incoming radiation, as absorption and emission of quanta cannot occur in zero time.) Thus it will be seen that our Milky Way is responsible for literally imprinting a specific degree of time retardation (or redshift) upon all extragalactic radiation.

This interpretation can be put to an interesting test. Upon the basis of theory, radiation received from low mass galaxies should be somewhat more redshifted than from equidistant high mass systems —

simply because the strength of "repulsive" quanta emitted from the more massive galaxies will be greater. In turn, this stronger flow will tend to facilitate the propagation of outward bound radiation in the face of an opposing flow streaming from our own system. Analysis of the redshifts of nearby galaxy clusters affords excellent confirmation of this prediction. In the M31 (Local Group) and M81 cluster, both of which have been subjected to close scrutiny, all 21 companions (of relatively low mass) are found to possess higher redshifts than their more dominant cluster constituents! The chance of this being an accidental occurrence is so incredibly slim that it must lend strong support to the premise of a D/R factor.

There is good reason to suspect that a noted "quantization" of redshifts among members of such galaxy clusters as Abell 262 and the Coma Cluster, for example, is similarly rooted in effects of gravitational repulsion. To the puzzlement of observers, it has been discovered that different types of galaxies (lying at roughly the same distance) display noticeably different redshifts. Those of high luminosity and concentrated mass tend to possess lower redshifts than their less brilliant and less massive neighbors. It may now be suggested that this peculiarity affords further confirmation of the D/R factor at work. In essence, the overall flux of "repulsive" quanta (flowing from galaxies) will impart very distinctive patterns with regard to time of light propagation — hence to redshift.

Yet another cause of such seemingly "quantized" redshifts may be related to geometrical effects inherent in the distribution pattern of galaxies, as prescribed by the "New Cosmology." By reason of this hitherto unforeseen repulsive factor many galaxies will be induced to exhibit hollow globular structuring within the confines of an overall supercluster, as the D/R factor acts to push such members to separations consistent with age and mass. Viewed from afar, a "bubble-like" sphere may be expected to present three distinct instances of redshift clumping. In addition to those foreground systems attempting to approach us due to repulsion from within the cluster, so an opposing background swarm will be receding (at higher speed) from the opposite side of such an expanding "bubble" configuration. Yet another above average clumping of galactic images will take the form of a dichotomous right angle distribution that is intermediate in distance between these two extremes. Accordingly, a majority of galaxies will appear to portray characteristics of a "quantized" nature, as the factor of D/R repulsion is permitted rather graphic expression over distances measured in tens of millions of light-years.

The Enigma of Galactic Redshifts

A rather curious and hitherto unrecognized redshift pattern, which will be seen to have profound cosmological implications, involves galaxies of our own Local Supercluster. Handicapped by a false model of creation, and in urgent need of a proper grasp of both gravitation and radiation propagation, the scientific community has failed to recognize certain remarkable features which will become evident with little more than a cursory examination of galactic redshifts. Not only will a study of our Local Supercluster membership reveal numerous contradictions with regard to traditional views, but a proffered explanation is capable of lending strong support to several basic principles expounded in the "New Cosmology." There is also reason to believe that such a study will afford additional insight into our proximity to the nearby cosmic edge.

A most informative catalog of galaxy redshifts, depicting both distance and location, was compiled by astronomer R. Brent Tully. Published by Cambridge University Press (ISBN 0-521-35299-1) and entitled *Nearby Galaxies Catalog*, this comprehensive listing of some 2,367 galaxies forms the basis for a hitherto neglected study.

Proceeding upon premises inherent in the "New Cosmology," it is expedient to plot Local Supercluster galaxy redshifts as a function of distance and direction. Dividing the combined redshifts (in km/s) of a large sample group by their total distance (in LY), it is possible to make an interesting discovery regarding average implied velocity of recession per MLY of separation. With further subdivision into a number of bins relating to celestial location and distance, certain startling facts emerge which cast even more doubt as to the credibility of Big Bang philosophy.

Embracing a study of some 1,590 sample galaxies, essentials of this landmark analysis are summarized in Table 1. A mere glance is quite sufficient to disclose a truly astonishing pattern — one in which the average redshift (per MLY) is revealed to be considerably higher in the direction of the cosmic edge. (Indeed, such redshifts are seen to be fully 31% higher than those of galaxies occupying an identical volume of space in the opposite direction of the sky!) The overall picture is unmistakable and cannot possibly be negated by unfounded arguments of limited statistics. There is clearly a definite and symmetrical decline of galactic redshifts (per MLY) from the vicinity of the nearby cosmic edge toward the center of the universe.

One other truly remarkable observation is that of a pronounced reduction of "center group" redshifts from the "under 50 MLY" bin to the "50 to 100 MLY" bin. Involving quantities of 153 and 405 galaxies, respectively, the difference between these two adjoining bins is an amazing 55%! Needless to say, this sharp decline of redshifts is a fact demanding clarification.

By coincidence, the cosmic center is seen to lie in the very same quadrant of the sky as the Virgo Cluster, a dense concentration of galaxies at the heart of our Local Supercluster. Thus it may be argued that gravitation acts to impose a measure of "infall" upon surrounding systems. But while this might conceivably enhance the redshifts of some galaxies lying between us and the M87 Virgo group, and to reduce the redshifts of others located behind this great cluster, gravitational attraction cannot be the answer. This conviction is borne out by the fact that most of the sample galaxies — between our Milky Way and the nearby cosmic edge — have quite similar redshifts (per MLY) to those residing between ourselves and the Virgo Cluster. Furthermore, as clearly shown in Table 2, adjoining bins (extending 20 MLY on either side of the Virgo group) are affected differently! The bin closest to us is 9.74 km/s higher; while the opposing bin is only 2.87 km/s lower. If the force of gravitation is responsible, both bins should vary by the same amount. Obviously, a more subtle factor is involved and we must look elsewhere for a full explanation.

But problems raised by this redshift study go beyond peculiarities of implied motion and the need to consider revised views of gravitation and radiation propagation. Far deeper cosmological issues are surely involved. Since the Hubble flow rate must exceed any real recession velocity among the galaxies of our Local Supercluster, due to mutual attraction among the constituent members, a serious contradiction to Big Bang philosophy will be seen to surface. The very fact that so many galactic redshifts are in excess of 25 km/s (per MLY) is strongly indicative of a Hubble flow of at least 40 km/s (per MLY). However, such a high rate would imply a Big Bang universe with an age/radius of no more than 7.5 BLY — a figure much less than the oldest stars residing in the globular clusters of our Milky Way! The alternative, of course, is to simply acknowledge either a different cosmological model or a non-velocity component of many galaxy redshifts — *or both!*

In order to place matters into true perspective, it is advisable to ponder the origin of nearby galactic redshifts, as a mystery of sorts is seen to beg an answer. Specifically, why do so many nearby galaxies exhibit a redshift? How did they ever manage to acquire a redshift if the D/R factor is ineffective at a distance less than about 17 MLY, which is the point where repulsion is predicted to merely offset attraction? If held together by the force of gravity, one should expect that the random motions of nearby cluster members would serve to guarantee a roughly equal mixture of blueshifts and redshifts. Only when it is finally understood how an inordinate number of redshifts were produced may we properly address the intriguing problem of such galactic peculiarities.

The preponderance of redshifts over blueshifts is revealed to be positively staggering. Of the 2,367 galaxies listed in the *Nearby Galaxies Catalog*, only 19 are seen to possess a blueshift! In point of fact, 9 of these are due to rapid orbital motion about super-massive M87 and several large companion systems. (By way of confirmation, it is noted that these approaching galaxies are offset by an equal number of exceptionally high redshifted and receding systems.) The

10 remaining blueshifted galaxies are all situated within a few MLY of Earth and tend to be strongly concentrated toward the nearby cosmic edge. In the face of this exceedingly biased redshift/blueshift distribution, it is little short of obvious that a *non-velocity factor must be deeply involved in the vast majority of all Local Supercluster galaxy redshifts!* Indeed, such a conclusion is surely beyond question; it only remains to advance a credible theory to account for this observation.

In all probability, a solution is to be found in the interaction of radiation with the myriad swarms of D/R quanta so believed to permeate space. More specifically, it is the ability of our Milky Way to retard incoming radiation — through ejection of "repulsive" quanta — that is largely responsible for nearby galactic redshifts. (It has already been firmly established that, in every instance where two equidistant galaxies are of widely differing mass, it is always the smaller system which displays the higher redshift. In turn, this must lead one to conclude that a more massive galaxy is better able to facilitate the propagation of expelled radiation — presumably, through ejection of a stronger flow of D/R quanta.) This being the case, it also follows that all galaxies possess an inherent ability to retard or slow incoming radiation — in the sense that the more massive the system, the greater the retardation.

Other than enhancement of redshift due to "cosmic arcing," and of interaction with rich clouds of interstellar hydrogen gas, it is logical to infer that the origin of non-velocity redshifts is the outcome of two opposing influences. On the one hand, the propagation of radiation — between galaxies — is assisted by the flow of D/R quanta streaming outward from the emitting galactic mass. Conversely, all galaxies must act to slow approaching radiation through ejection of this same form of "repulsive" quanta. However, it is important to realize that the chief influence is bound to be the one imposed upon any incoming radiation. (This assumption is rooted in a combination of large-scale scattering of D/R quanta over distance and the small-scale "focusing" of such quanta as it is condensed in proximity to a massive galaxy.) Hence, the noted "surplus" of galactic redshifts — toward the cosmic edge — may well have a ready explanation in the premise of extinction at the boundary of the universe, which acts to severely restrict the flux of inward flowing "repulsive" quanta. As a result, a superior outward flowing stream is enabled to impose a degree of retardation, or time delay. In the opposite direction, a stronger flow of "repulsive" quanta will permit our Milky Way to reflect this energy back toward the cosmic center (much as a mirror reflects light), which would account for the above average redshifts in the centrally located "under 50 MLY" group.

While it is tempting to ascribe the sharp reduction of center group redshifts (beyond 50 MLY) solely to infall toward the Virgo Cluster, a major factor is seen to be the strong flow of D/R quanta ejected from this same massive cluster, which acts to push radiation along with increased efficiency. Inasmuch as any retardation of radiation hinges upon both field strength and transit time, our Milky Way is less able to impart a non-velocity redshift upon more remote sources by reason of dissipation of D/R flux with distance. Hence, in conjunction with a reduction of flux due to expansion of the universe, systems located immediately beyond the Virgo Cluster may be expected to exhibit even lower redshifts (per MLY) until finally stabilizing at a rate more commensurate with the dictates of Hubble's law— exactly as observed. (In the direction of the cosmic edge, where no such massive clumping of nearby galaxies exists, a similar abrupt redshift reduction — beyond 50 MLY — is precluded by a scenario featuring a somewhat more homogeneous flow of D/R quanta.)

To the astonishment of the astronomical community, it has been reported that the redshifts of certain members of our Local Supercluster are actually declining with the passage of time! [5] Upon analysis of observational data, compiled during the course of the past decade or two, it is claimed that (in this interval) a reduction of typically some 1 to 2 km/s is probably the rule rather than the exception. Needless to say, *upon the basis of traditional scientific beliefs, there is no conceivable explanation for what could prove to be a truly bizarre state of affairs!*

In light of theorems inherent in the "New Cosmology," it may well be instructive to examine certain peculiarities which must arise in the vicinity of the nearby edge, where extinction serves to severely restrict the flux of "repulsive" quanta flowing back toward the cosmic center. In turn, this negation of D/R quanta will allow any inbound radiation to escape the normal repulsion exerted by our Milky Way, thus permitting inbound corpuscles to reach us in less time and with reduced redshift. (No perceptible change is to be expected in the spectra of very remote sources, as light transit times become overwhelmingly large in comparison to any reduction.)

The question which now comes to mind is whether nearby redshifts could actually turn into blueshifts. Clearly, the prospect of just such an occurrence, in the ridiculously brief span of only a few thousand years or so, is enough to raise eyebrows. This issue is likely to be at the cutting edge of future research and debate as astronomers, physicists and cosmologists desperately seek to resolve a paradox which is greatly compounded by erroneous views of gravitation and radiation propagation — to say nothing of a false creation model!

By way of coming to grips with the problem, it is advisable to ask one pertinent question. In short, how is it possible for non-velocity redshifts, which take millions of years to arise, to be negated in but a small fraction of this time? Clearly, an explanation requires some sort of mechanism capable of producing a sudden and drastic impact upon radiation as it enters our Milky Way's sphere of influence. It must be evident, in fact, that traditional laws of physics are quite inadequate to resolve what (in the eyes of many scientists) must pose a puzzle to end all puzzles.

In an encouraging test of "New Cosmology" theory, a simple and most logical solution comes to mind. It has hitherto been presumed that both radiation and gravitational energy (including D/R quanta) are propagated at essentially the same velocity — namely, the speed of light. But what if "repulsive" quanta travel slightly faster than radiation? (The very premise that gravitation can flow from a black hole, while radiation cannot, is consistent with this hypothesis!) A difference of only a fraction of one percent can be shown to have a significant impact upon radiation that has been in transit for millions of years. Moreover, should D/R quanta be slightly less affected by cosmic arcing, this repulsive force will reach us by a more direct route and in less time. Together, this ability of "repulsive" quanta to outrun radiation (which is constantly having its speed adjusted through interaction with the infinitesimal quanta of space) will lead to the spectacle of a rapid decline of redshifts with respect to an Earth observer residing in the immediate vicinity of the cosmic edge.

To grasp the full picture of how such differential velocity could cause a sharp reduction of nearby galactic redshifts, it is well to ponder matters in conjunction with the annihilation factor that is believed to await us at the boundary of the universe. Since the last D/R quanta (emitted from a long-extinguished galaxy) will be seen to have passed us before arrival of its last radiation, it follows that the normal ability of the Milky Way to retard incoming radiation will dwindle rather abruptly — simply because our system will no longer have as much "repulsive" quanta to eject toward such inbound corpuscles. As a direct consequence of this sudden cutoff of the means with which to slow approaching radiation, a situation must prevail whereby the non-velocity redshifts of many local systems are reduced by reason of extinction at the cosmic periphery. Conceivably, a detailed study of the rate of change of neighboring galactic redshift sources may one day afford a new and totally independent method of determining our proximity to this most portentous boundary — an opportunity which hinges upon accurate assessment of an inferred transit time differential between radiation and D/R quanta.

This reported decline of nearby galaxy redshifts is exactly the proof needed to confirm essentials of the "New Cosmology," including the prediction that we presently reside in close proximity to an exceedingly fateful cosmic edge! Nature has seen fit to provide us with valuable clues concerning our status and location in a truly dynamic universe. It is to be hoped that the scientific community will not permit the sin of prejudice to cloud judgement, nor restrict the quest for knowledge should this search threaten the credibility of established views. Most assuredly, any decline of local galaxy redshifts positively demands an urgent and unbiased investigation in the light of viable new principles.

The Microwave Background

In the opinion of many scientists, the observed microwave background radiation originated some billions of years ago in the intense heat and density of a Big Bang explosion. It is believed that expansion of the universe has served to reduce highly energetic radiation to the noted wavelength peak of about 1 mm which, in turn, has been translated to a temperature of close to 3° K. But this concept is highly illogical; in fact, it bespeaks an uncritical assessment of current nebulous views of radiation.

The very idea of stretching wavelengths — from the realm of high-energy gamma rays into the region of enormously weaker microwaves — poses a vexing contradiction with a basic tenet of physics demanding conservation of mass-energy. Ignoring the issue of particle mass, it fails to explain how such entities as electron/positron combinations (gamma rays) can incur a drastic reduction of mass simply because the universe is expanding. (Upon the basis of the "New Cosmology" every component of creation is seen to possess the property of mass, or ability to commune gravitationally — a property which may be negated, as in the case of the elusive neutrino and radiation in general, solely due to extreme relative motion at the speed of light, since this is essentially the velocity by which gravitation is propagated.) It is thus inconceivable that cosmic expansion can have any effect upon the actual mass-energy congealed into corpuscles of radiation. All that it can do is to impose a reduction of flux and of frequency (wave crests per unit time) by reason of Doppler shift. However produced, the microwave background radiation cannot undergo any substantial change (from time of formation) by reason of an expanding universe.

The low energy level and the remarkably isotropic nature of this phenomenon is suggestive of an entirely different origin than a Big Bang explosion. What is surely indicated is an homogeneous production mechanism, more characteristic of the near *absolute zero* of outer space, than of the incredibly high temperatures of a primordial fireball. The most probable scenario to emerge involves elementary matter forms that are strewn throughout the expanse of space, where the chief source of energy is the larger "repulsive" quanta so believed responsible for expansion of the universe. The problem is one of explaining how exceedingly minute quanta are absorbed, and then emitted as more massive microwave radiation. In essence, *the microwave background reflects the average temperature of matter in the cold isolation of outer space!*

While a great deal remains to be explored in depth, it is logical to infer that (in the immediate vicinity of "maximum" particles) quanta of a certain size range will infrequently exist in mutual communion or association long enough to enter into a degree of fusion. When the inevitable expulsion occurs, it could well be in the form of discrete corpuscles with masses characteristic of the microwave region of the spectrum. (It is to be expected that expulsion of absorbed quanta, in the almost absolute zero of space, will not be so abrupt as would be the case at a higher temperature.) Although only a small fraction of ingested quanta is likely to congeal into such radiation, what is produced will be propagated back and forth among components of the interstellar medium. Thus a universal sea of microwave radiation (which just happens to be centered about the 1-mm wavelength) may be presumed to arise by reason of creation in the halos of isolated material particles, rather than originating in a superdense explosion.

While the vast majority of Big Bang proponents still cling to the belief of a fireball microwave origin, a growing number of reputable scientists have succeeded in presenting an impressive argument to the contrary. In spite of widespread refusal to recognize the idea of low wavelength radiation scattering due to interaction with the extragalactic medium, there is strong evidence that this is indeed the case. [6] Not only are radio/microwave corpuscles effectively scattered during the course of a lengthy transit, to produce an ubiquitous celestial fog of low-level radiation, but this inferred dissipation of quasar radio emissions readily explains an otherwise mysterious decline of such flux which is observed to occur with distance. (It is known that, in general, radio-loud quasars are relatively nearby and of more modest redshift.)

Since passage through a uniform field of microwave radiation is bound to produce a Doppler shift, measurements have been made to determine our direction of motion through space. It is reported that the microwave temperature is slightly higher in the direction of the constellation of Leo; while it is less by an identical amount in the opposite direction of the sky in the constellation of Aquarius. This has been interpreted to mean that the Milky Way, along with other members of our Local Group, shares a common motion (about 600 km/s) toward Leo. On the surface, this would seem to conflict with evidence that our entire region of space is moving toward Sculptor. What may be wondered is why observation appears to indicate movement in almost the opposite direction!

To resolve this paradox it is most essential to contemplate certain aspects of cosmic expansion in the light of the "New Cosmology." Since absorption and emission of quanta cannot occur in zero time, it is logical to infer that a massive galaxy will resist acceleration more than minute corpuscles. Indeed, in order to impart motion to vast star systems, it is necessary for the impetus of "repulsive" quanta to be transferred from atom to atom — a process which is somewhat less than instantaneous, as galaxy components are held together by gravitation. It must follow that this time lag will enable all outward-bound radiation to have a slight edge over similar corpuscles propagating toward us from the direction of the nearby edge. In effect, any lag in our Milky Way's outward acceleration is tantamount to reducing the Doppler shift of incoming quanta (from the vicinity of the cosmic center) — thus explaining why the background radiation is observed to be a trifle warmer toward Leo.

The small discrepancy between microwave observations and quasar distribution, as a means of determining our orientation with respect to the cosmic center and nearby edge, would appear to have a ready answer. In short, it is likely due to residual orbital momentum stemming from the Small Bang which gave us birth..

In negating this last remnant of support for a Big Bang cosmology, it is clear that evidence against this model is now so overwhelming that a parallel exists with the state of affairs which prevailed during the pioneering days of Copernicus, Kepler and Galileo. It is to be hoped that science will soon recognize the futility of attempting to preserve an outworn concept and begin to explore new horizons. The reward will surely justify the effort.

References:

[1] See: "A New Non-Doppler Redshift," by Ph.D physicist Paul Marmet *(Physics Essays, Vol. 1, pps. 24-32, 1988)* This very principle is also expounded in another paper by Paul Marmet, entitled "Red Shift of Spectral Lines in the Sun's Chromosphere" (IEEE Transactions on Plasma Science, Vol. 17, No 2 , April 1989).

[2] An informative treatise of discordant redshifts may be found in a book by the eminent astronomer, Halton C. Arp. This truly monumental work is entitled *Quasars, Redshifts and Controversies* (ISBN 0-941325-00-8).

[3] See: "Hubble Space Telescope," *Astronomy*, February 2000, page 24.

[4] W.G. Tifft, and W.J. Cooke, "Quantized Galaxy Redshifts," *Sky & Telescope, January, 1987.*

[5] W.G. Tifft, *The Astrophysical Journal*, 382; pps. 396-415, December 1, 1991.

[6] See: *The Big Bang Never Happened* (p. 277), by Eric J. Lerner (ISBN 0-8129-1853-3). See also the very enlightening paper, *An Alternative Interpretation of the 3 K Radiation*, presented by a leading theoretical physicist, Professor Paul Marmet ("The American Association for the Advancement of Science," San Francisco, June 1994).

Note:

Although not generally realized, it must be stated that there exists today an almost unbelievable degree of bias in astronomy and physics— with strong overtones extending into cosmology. Regrettably, it has now reached the point where many budding scientists are actually afraid to voice criticism of a Big Bang creation, or of any facet of Einstein's relativity theory. To do so is to risk one's career, as critics are promptly ostracized and few (if any) science journals will accept their papers for publication. Moreover, dissenters have little hope of obtaining a research or teaching position. It is the shame of our society that true science is thwarted by a governing clique of so-called "authorities" who fear loss of prestige should their views be overturned. One can only hope that this unjustified prejudice will soon give way to an atmosphere of genuine science, in which there is a willingness to examine evidence in an impartial manner.

6

Science and Philosophy

The Great Cosmic Pyramid — The "Time Constant" of Fusion — Beyond Human Evolution — A Cosmic Paradox — Summary

Both theoretical and observational evidence would appear to lend strong support to the concept of a bound and finite physical universe. But many questions remain unanswered. Not the least of our problems is the need to devise a formula linking cosmic expansion with the *Motivation* behind creation. Furthermore, it must be conceded that the very existence of an enclosed system does not preclude the possibility of *other* universes! (Might our universe be as but one cell in some overall body?) However, restricted as we are by limited insight, it is perhaps prudent to reserve comment to that which is conducive to investigation. In so doing, it will be shown that the physical cosmos is characterized by a rather unique series of mathematical relationships — a feature which would seem to indicate that, for all practical purposes, *our own universe must constitute a bound system that is sufficient unto itself!*

The Great Cosmic Pyramid

A most instructive picture of the universe emerges as a result of scientific deduction and logic. At the lower extremities of the Cosmic Pyramid, the intrinsic worth or value of a particle of matter is revealed to be directly related to the quantity of "congealed energy" fused together into a common body — an entity that is as one with regard to such properties as space and time. Above the realm of atomic nuclei, with their enormously enhanced complexity and status in comparison to an elementary "**A**," we soon encounter a revolutionary stage of expression in the form of organic life.

This new level is characterized by a marked transformation, in the sense that fusion of larger quantities of "**A**" no longer serves to produce a corresponding increase in the mass of an entity. At this point any further "desire" must be manifested in some other manner and upon an entirely higher plane. For instead of expending energy as the phenomenon of gravitation, in an endeavor to attract mere matter which now lies well below its status, such effort is directed to the task of achieving maximum interaction and harmony with similar members of its own world. Subsequently, as organic evolution progresses upward to the status of man at his pinnacle, so it may be inferred that these higher levels of spirit must represent enormous amounts of energy since contained and converted to increased segments of God's Nature. A superior cosmic status, being the product of *deserved* spiritual fusion, may only be obtained with the passage of time and at the expense of a great deal of effort or energy, as we are often quite aware.

Even the strange paradox of Eternity begins to shed some of its aura of mystery once the phenomenon of time receives interpretation as a relative experience — one that will surely vanish among members should interaction ever reach the point where it becomes instantaneous. To the extent that a critical degree of reciprocal communion is achieved, so there will no longer exist any reason for separate identities, and the outcome will be a loss of previous personalities as they become united into fewer entities of higher worth. Time may thus be viewed as a fundamental attribute of *imperfection*, prevailing at all levels of creation below that of God, but having no substance or meaning once Perfection is reached — a state which might well be described as one of *Total Communion or Instantaneous Motion.* Eternity or Timelessness, rather than the transitory world of space and time, is disclosed to be the Ultimate cosmic state. Just as surely as time appears at the lowest strata of creation, so it will cease to have any meaning at the apex of the Great Cosmic Pyramid once spirit is deserving of being united into One Harmonious Whole.

Viewed upon the large scale, the rate by which spiritual fusion occurs throughout the universe cannot possibly be ascribed to a random factor. It is clearly a phenomenon in which advancement at one level is intimately connected with progress achieved at all other levels. The equilibrium of the cosmos can only be maintained if spirit of the lowest caliber is created at a rate capable of balancing evolutionary movement to higher levels. Similarly, frequency of fusion must be such as to facilitate a steady upward flow of spirit, free from the catastrophe of disruption that would arise should either a shortage or surplus be produced at any one level.

Notwithstanding this necessity to preserve overall stability, it is important to realize that — on a short-term basis — there need not be an ironclad rule insisting that actual instances of physical fusion must always take place, over the entire range of cosmic "steps," with unswerving regularity. On the contrary, it is sufficient only that adjustments intercede periodically to restore long-term balance and justification. Judging by the relative abundance of the many physical body-forms, housing a wide range of spiritual complexes, it is evident that phases of both literal fusion and quasi-fusion exist side-by-side in the universe.

Although a pattern of literal fusion (with reduction of numbers) may be inferred during the evolution of "**A**" to the level of atomic particles, a certain complexity surrounds the involvement of animate structures. A vast gap seems to prevail between the numbers of the material particles and elementary life forms. This discrepancy must surely indicate that, upon occasion, a great many spiritual entities may undergo delayed fusion into one radically different and higher structure. Prior to this significant event, justice could be rendered in a quasi-fusion state. In a sense, spirit may enjoy the "fruits of fusion" without a literal fusion having transpired. At the "pre-life" stage this condition is probably fulfilled within the extremely hot and dense interiors of the stars, especially degenerate stars where acts of reciprocal communion must be so magnified as to simulate actual fusion.

The course of biological life appears to be characterized by a mixture of fusion and quasi-fusion. In addition to a reduction of body-forms with status, there are also instances where the advantages of fusion — without fusion — are readily available to many life forms. Man, for example, affords an excellent illustration of the benefits accruing from the ability to commune with other members of one's own species. By reason of this opportunity to cooperate and share in a wealth of knowledge and experiences, his environmental status is seen to be enhanced to an extent not otherwise possible.

This leads us to ponder the long-term prospectus of mankind as a physical entity. Quite regardless of any inferred future reduction of population with evolutionary progress, it is inevitable that man's present quasi-fusion era must continue to alleviate instances of any spiritual injustice until his planet's final demise. What then? Indeed, since it is rather inconceivable that man as such could ever expect to conclude his evolution toward Perfection without there being some intermediate stage, we may well ask the nature of this new and highly revolutionary structural form.

Strangely enough, his destiny could very conceivably be found in the stars! If this should seem slightly melodramatic, we have merely to consider the unique properties of fully degenerate stars — stupendous concentrations of matter so compressed as to virtually constitute *super atomic particles!* Not only can we point to the incredible density of a neutron star (about half a million Earths compacted into a sphere only 10 miles or so in diameter), but the stellar black hole state of matter may be described as a total collapse and absolute fusion of a slightly greater segment of creation.

Becoming manifest as a spiritual entity of tremendous worth, once a critical stage of contraction is achieved, yet another quasi-fusion era may be initiated as a star of superdense proportions continues its development upon a new plane of existence. One can only speculate as to the details involved in subsequent evolution. Presumably, progress is facilitated by means of establishing a degree of reciprocal communion with similar entities — *interaction of a non-physical nature and no longer restricted to the velocity of light.* Finally, upon attaining the ability to express Instantaneous or Total Communion, time will cease to exist for such advanced entities and a *literal fusion into God will ensue!* Somehow, as a consequence of *black holes* concluding this momentous transition from the physical world of imperfection and time, the surplus "energy" so released is manifested as a new generation of "**A**" within the confines of a *Cosmic Creation Zone.*

By way of illustrating the rather close interrelationship that exists between certain cosmic extremes, we might point to five rather curious numerical associations embracing phenomena of the ultra-small and the ultra-large. Involving the truly immense sum of 10^{40} (1 followed by 40 zeros), these remarkable similarities may be stated as follows:

(1) The electrical charge of a positive/negative pair of material particles exceeds their mutual gravitational influence by a factor of very nearly 10^{40}.

(2) The density ratio between highly degenerate stars and the average density of matter throughout the expanse of the universe is of the order of 10^{40}.

(3) Stretched out in a straight line, like a string of beads, it would take about 10^{40} proton nuclei to reach the distance at which the galaxies are receding at the speed of light.

(4) The repulsive aspect of gravitation, at a distance equal to the

diameter of a "maximum" particle nucleus, is seen to be some 10^{20} times weaker than attraction — in essence, strength of repulsion is the square root of attraction.

(5) The square root of the total number of protons, within the expanse of our "observable universe" (10^{80}), is none other than 10^{40}.

In each instance these relationships may apparently be looked upon as an expression of fusion versus non-fusion. Not only do they tend to substantiate the underlying cosmic principle of fusion, but there is even reason to suspect that *a valuable clue lies therein as to the number of "A"s fused within the level of a "maximum" particle of matter!*

It will be recalled that, in the final analysis, the phenomenon of gravitation might be defined as an inborn desire for fusion on the part of all material components of creation — a universal expression of intrinsic worth dissipated over the depths of space. It is essentially a reflection of a particle's ability to achieve attraction by means of communion *with the physical universe as a whole.* In contrast, the electrical force of a particle is a manifestation of this same worth condensed to a volume of space equal to that of its *immediate self* — namely, the halo of energy surrounding the material body in question. Since there is good reason to believe that the value of "**A**" fused within a proton or an electron does not exceed that of the halo by a really exorbitant figure, it follows that exposure to another charged particle will produce a magnified reaction that is *proportional to the quantity of "A" involved!* Hence, it is logical to infer that a "maximum" particle likely consists of roughly 10^{40} "**A**"s. In a sense, the electrical force may well be considered an indication of achieved fusion; whereas gravitation may be viewed as a sure sign of potential fusion associated with Destiny.

With regard to the second relationship, a fully degenerate star must represent a state of communion tantamount to "maximum" density or fusion into one overall entity. On the other hand, the average density of matter in space may be defined as the "minimum" degree of interaction that is permitted to exist without requiring new creation. Within the Cosmic Creation Zone, the introduction of "**A**" will be inferred in all regions falling below average density, in contrast to those forms that are able to achieve critical condensation leading to a degree of total fusion. In essence, these extremes may be thought of as depicting maximum and minimum concentrations of energy within the framework of a bound system — one in which the velocity of light denotes a total separation among components of creation.

The connection between the diameter of a proton nucleus and the size of the universe is, of course, directly linked to the issue of average density and hydrogen abundance. Starting with the fused state of such a particle, it will be seen that this overall distribution is consistent with a twofold increase of distance (an eightfold increase in volume) for each fourfold increase in the number of proton nuclei. Restated slightly, the pattern is one of a fourfold decrease in cosmic density with each doubling in the quantity of matter residing within the universe. To extend the picture, it is noted that a fourfold increase in mass is capable of expressing the same strength of any gravitational communion over eight times the volume of space. On the surface, this is possibly surprising, since it might be wondered why the two should not coincide. What mysterious factor enables the force of gravitation to permeate as much space as it does?

The explanation presumably hinges upon a proper assessment of the roles played by time and evolutionary processes. For every increase in the creation of matter, so must it add to the task of achieving full communion in a universe where interaction among material structures is regulated by the velocity of light. Should we proceed upon the premise that every level of spirit is able to double its status in unit time, it may be deduced that each successive level must possess the power to attract twice the volume of "**A**." But since a larger amount of matter must lengthen the time required to unite creation, such an addition can only take the form of increased distance or separation. The ensuing pattern of isolation, which would preserve equilibrium by fulfilling the basic requirement of each "**A**" being able to obtain fusion with another "**A**" in the same interval of time, is one of a fourfold increase in matter with every doubling of distance. Upon this premise a twofold decrease of density — beyond what would be expected to characterize a simple doubling in the quantity of matter — will be seen to accompany each "step" in the Cosmic Pyramid. For not only must spirit be unsatisfied to maintain the status quo, but it is motivated to seek its Ultimate Destiny through further fusion. Hence, in pondering the ability of gravitation to spread its influence over space, we are very likely witnessing a graphic demonstration of spirit striving unceasingly to effect a *future* doubling of status.

A philosophical basis for the inferred D/R factor would thus seem implicit in theory. It is surely more than a coincidence that, at the point where this repulsive force first becomes manifest, it does so as the square root of the number of fused "**A**"s which are now believed to constitute the level of a "maximum" elementary particle. Indeed, it is essential that a twofold increase in the strength of gravitational repulsion induce a fourfold increase of distance in order to achieve a balance between the rate of new creation and the rate of cosmic evolution. *In effect, it is the rate of spiritual advancement which determines the time scale of Hubble's constant!*

With a total of 10^{80} protons comprising our physical universe, and 10^{40} fused "**A**"s per proton, it will be deduced that Perfection — or any one "level" within the Great Cosmic Pyramid — is equivalent to some 10^{120} "**A**"s. In terms of spiritual evolution, *this Ultimate State of Perfection would involve approximately 400 "steps" or occasions of doubling.* Subsequently, an entity of the status of a proton is computed to reside 1/3rd of the way toward the top of the Pyramid, having already climbed some 133 "steps" from its origin as an infinitesimal "**A**." Upon this premise, slightly less than 1% of all the mass-energy in the universe is likely to be incorporated into actual "maximum" particles such as the proton.

This assumption can be put to an interesting test. What may prove to be so highly significant about this figure is the excellent agreement with certain cosmological considerations as they pertain to the Steady-state model of creation. According to this principle there are very definite limitations as to the relationship between average density of matter in space and the rate of cosmic expansion. With Hubble's constant established at roughly 12 billion years, no more than a few percent of all the material in the universe may be so concentrated as to form galaxies without threatening stability. Were a substantial proportion locked up inside stars it would contradict the assumption of Continuous Creation, which demands a preponderance of newly created matter forms in space. Conversely, too great an abundance between star systems — with Hubble's constant being what it is — could lead to the spectacle of our gravitationally weak Local Group of galaxies possessing a lower than average density — thereby causing the cluster to expand apart, which it is not doing. On the whole, our inferred ratio of mass-energy to actual matter, within the expanse of the universe, is seen to afford a most remarkable agreement with observation.

The "Time Constant" of Fusion

Certainly a most intriguing prospect now awaiting us is that of determining the universal *"time constant,"* which must surely regulate the entire process of spiritual fusion. For it is an integral feature of theory that all levels of spirit, within the Great Cosmic Pyramid, must succeed in doubling their worth in some common interval of time. A solution to this problem will make it possible to derive a useful picture as to the average time that an individual must serve in order to achieve the equivalent of one doubling of status. Moreover, it will also provide insight into the highly relevant question as to when one may expect a future rebirth.

Evaluation of this time aspect entails consideration of certain factors. For instance, the rate of disintegration (at the edge of the universe) must be included in any equation attempting to express events in terms of a human lifetime. Due to a steady annihilation of worlds containing a variety of life forms, there will prevail a situation in which disrupted levels of spirit are compelled to seek rebirth on appropriate planets still residing within the Cosmic Creation Zone. Along with other factors pointing toward a similar conclusion, this introduces the suspicion that, in terms of the planet with which a spirit was last associated, considerable time could elapse between reincarnations. In other words, biological evolution is likely to reflect a "sharing" of available body-forms with displaced spirit from alien worlds — with the result that spiritual entities are, in effect, excused from having a physical manifestation over most of the course of a planet's lengthy history! (Time, of course, simply does not exist to spirit between incarnations, just as it appears to cease when we fall into a deep and dreamless sleep.)

Thus, in theory, it may be presumed that those who manage to increase their intrinsic worth in excess of the average will receive what might well be equivalent to a longer postponement in their next rebirth — thereby enhancing prospects of securing an existence in more suitable surroundings. On the other hand, one who has made relatively little or no progress (whether or not restricted by premature death) may expect to be reborn in but a fraction of this time and, if so justified, in a less favorable environment should a regression be involved. Perhaps, in this manner, nature permits us some latitude in determining the circumstances into which we are reborn — including the specific moment of rebirth within an evolutionary era. Most assuredly, there is a very definite advantage to be gained from being reincarnated in a more enlightened and civilized age.

A truly great challenge, confronting the inquisitive mind of mortal man, is to deduce the *"time constant"* of spiritual fusion. For without a steady and orderly upward flow of spirit, toward an Ultimate State of Perfection at the apex of the Cosmic Pyramid, the universe would be totally lacking in both Purpose and Justice. In spite of the many uncertainties that must be encountered when pioneering into virgin territory, a solution of the time/evolution relationship is a task which cannot be avoided if true knowledge is to be our goal. While admittedly posing a structure of immense complexity, our universe is not without numerous clues and much vital information — information that would have been treasured by many wise scientists and philosophers of the past. If complete success is not to be within our immediate grasp, it should never be because we lack the fortitude to try.

Clearly, the *rate of expansion* of the universe must constitute a key factor in any approach to the time aspect of cosmic evolution. Why do galaxy superclusters double their distance from each other in an interval of close to 12 billion years? Why this particular figure? Obviously, the greater the rate of expansion, the greater the degree of disintegration at the edge of the universe. Inasmuch as this annihilation factor must be balanced through creation, it will be seen that the introduction and evolution of new "**A**" is closely related to the recession velocity of the galaxies. The issue, therefore, is basically one of determining precisely how "**A**" is introduced with respect to each cycle of Hubble's constant.

Upon due consideration, the one logical possibility would seem to embrace the premise that *one entire level of new "A" is created in just such a span of time and is able to fulfill its Ultimate Destiny by evolving to a state of Perfection in this same interval!* Should this be the case, it must follow that the time required by spirit to double its worth may be determined by dividing Hubble's constant by the number of "steps" or doublings separating "**A**" from Perfection.

Assuming a proton to consist of some 10^{40} "**As**," and a physical universe of 10^{80} protons, we have a spiritual pyramid of 10^{120} "**As**" per level and a total of approximately 400 "steps." (In view of the most unique relationship of the sum of 10^{40}, as it relates to the size of our finite physical universe, along with the inability of material forms to communicate gravitationally at velocities exceeding that of light, it is perhaps in order to treat such a region as though it constituted a complete and closed system terminating in what — for all practical purposes — we may define as God.) Subsequently, a period of roughly 30 million years per occasion of doubling may be derived.

Proceeding upon this basis, it will be deduced that only 1/400th of an entire level of new "**A**" will be created in a time span of some 30 million years. In effect, this means that only one spiritual entity in 400 may have a "conscious" (or physical) manifestation at any given instant. But if 30 million years is the allotted time, in which one entire level of spirit must double its status, this very restriction will introduce an additional factor of 400 into the time permitted for actual instances of physical incarnation. Dividing our 30 million year period by this further factor of 400, we are left with about 75,000 years as the average elapsed time for a "conscious" existence between doublings.

Yet another factor is imposed by disintegration at the edge of the universe, as displaced spirit must periodically be reborn within the confines of the Cosmic Creation Zone. Since it takes time for new stars to form and for planets to evolve biological entities, it is evident that this factor is very much related to status. Calculation of the average age of matter, within the framework of the Steady-state model of creation, yields the information that only about one part in 20 will be as old as Hubble's constant.

Upon this line of reasoning, the average "conscious" time that is required by spirit to double in value may expressed by the formula:

$$T = \frac{SH}{DLS^2} \quad \text{or} \quad T = \frac{H}{DLS}$$

Where:

T = Time required to achieve one doubling of status in terms of a "conscious" existence.

S = Total number of "steps" from "A" to Perfection (or Full Communion) within the confines of a physical universe that is some 18 BLY in radius.

H = Hubble's constant (deduced to be about 12 BY).

D = Disintegration factor at the edge of the physical universe in one cycle of Hubble's constant.

L = Level of spirit in question — defined in terms of "steps" evolved beyond the initial state of "A."

Assuming that 400 "steps" extend between "A" and a state of Total Communion, it now becomes essential to determine the position of mankind in this overall scheme. With a hydrogen nucleus established at "step" 133, this problem may be resolved by a comparison of the relative abundance of the proton to man throughout the Cosmic Creation Zone.

By way of assessing this ratio, it may be estimated that about 10^{22} Earth-type planets presently exist in the universe, of which roughly one in 10^5 will currently feature biological evolution equal to man at his present level. Multiplying our estimated 10^{17} planets by the sum of 10^{10} (representing an average long-term population of man per planet), we arrive at a total of 10^{27} human beings at this one particular level. Since some 10^{80} protons constitute our physical universe, we may deduce a ratio of about 10^{53} protons for every man. Translated into "steps" in the Cosmic Pyramid this sum is equal to some 177 doublings. Accordingly, the level of man may be derived by adding these 177 "steps" to the 133 "steps" believed to characterize a proton — leading to the conviction that mankind presently resides at about "step" number 310.

With this information, we may now attempt to resolve the "time constant" for man as per our formula. Therefore:

$$T = \frac{12{,}000{,}000{,}000}{20 \times 310 \times 400} \quad \text{or} \quad 4{,}800 \text{ years (approximately)}$$

Where:

T = Time required to achieve one doubling of status in terms of a "conscious" existence.

Thus, it would be calculated that, upon the average, man must serve a total of something like 4,800 years (or almost 70 lifetimes of 70 years) in order to double his intrinsic worth. (Any uncertainty in computing the ratio of man-to-proton will not seriously alter our estimate, since an error by a factor of a million will only make a difference of about 6%, and a million million error only a 12% difference!)

With such insight into mankind's status in the Cosmic Pyramid, along with knowledge of the "time constant" of spiritual fusion, we may now face the issue of reincarnation in terms of Earth evolution. During what past age did we last live? Moreover, in what future era are we likely to be reborn?

Considering the necessity of having to accommodate displaced spirit from other worlds, along with the overall cycle of doubling which is of the order of some 30 million years, reincarnation may essentially take the form of a projection well into the future and into quite another planetary environment. Thus, if man requires 4,800 years as a "conscious" existence to achieve one doubling of status over a period of 30 million years, it will be tantamount to an elapsed time of almost 6,200 years for each year of a past life. Based upon an average life expectancy of 70 years this would call for a rebirth some four or five hundred thousand years into the future. However, if only 1/400th of an entire level of "**A**" is created per 30 million year cycle, in terms of Earth evolution the *apparent* elapsed time will be less than the *actual* time by this very same factor. Subsequently, for all intents and purposes, the reincarnation of man may be linked to the basic formula:

$$R = \frac{YH}{TS^2}$$

The two additional symbols may be defined as follows:
R = Rebirth of man (in years) in terms of a future era on Earth.
Y = The number of years lived during last incarnation.

$$\text{Hence: } R = \frac{70 \times 12{,}000{,}000{,}000}{4{,}800 \times 400 \times 400} = 1{,}100 \text{ years (approximately)}$$

In other words, what probably transpires is that man is actually reborn some hundreds of thousands of years from now, but in relation to the environmental status with which he was last associated it will be equivalent to an evolutionary era only a thousand years or so into the future.

Beyond Human Evolution

Upon contemplating the future of man, as a biological species, it may be theorized that each additional doubling will require some

75,000 years in terms of Earth evolution. The eventual passage of mankind, beyond his present transitory and rather frail human form, could very well have a solution in the astronomical curiosity of a *celestial black hole!* (Neutron stars, which may in time evolve into stellar black holes as gradually declining rotation permits still further contraction, could also share such an exalted status.) Best described as immense concentrations of mass-energy contracted into a state bordering upon spatial oblivion, the ratio of mankind to this unique state of fused matter should allow computation of the number of "steps" needed to achieve the inferred evolution. If this is indeed the case, it may in truth be said that man's destiny is in the stars!

Strange as it might first appear, there are some very convincing arguments for linking man's future to a phenomenon that would tend to tax the imagination. Not only does this compact state of "maximum" density meet all the required attributes of fusion — thereby constituting a spiritual entity of a specific level — but an excellent agreement of numbers can be derived which would connect theory with observation. It is surely more than a mere coincidence that an estimate of all the stellar black holes, likely to exist in the universe, should turn out to be equal to the probable number of Earth-type planets that are permitted to fulfill their evolutionary roles. Moreover, it can be shown that the quantity of "**A**" fused within a typical stellar black hole (some 10^{58} protons x 10^{40} "**A**s," or about 10^{98} fused "**A**s"), when multiplied by the number of such objects in the entire universe, is very likely of the same order as the total spiritual worth (or "**A**" content) of one complete population of Earth-man multiplied by the quantity of all other worlds of a similar evolutionary level. The implication is little short of obvious: *A stellar black hole may well represent the eventual destiny of a humanity, so deserving of intimate harmony, as to become fused into a new entity of truly enormous magnitude!*

By way of determining the number of cosmic "steps," which exist between present-day man and the lofty status of a stellar black hole, it is necessary to examine what transpires as man finally evolves beyond his current level, which we might define as M^1. Upon doubling his worth, and attaining the status of M^2, it will be noted that a gap has suddenly appeared in our planet's otherwise unbroken chain of evolutionary life forms. In effect, there is no longer an M^1 level. Similarly, after M^2 progresses to the status of M^3, the gap will widen to include the absence of both the M^1 and M^2 levels. For the first time in our planet's history *a gap has occurred in the very evolutionary continuum* — one that will grow steadily in size until man's eventual extinction as a biological entity. Assuming that only one Earth-type planet in 10^5 will feature human life equivalent to a specific "M" level, it may be computed that about 17 doublings will suffice to accomplish the inferred evolution from M^1 to the status of a stellar black hole. Accordingly, this momentous stage may be looked upon as constituting "step" number 327 in a Cosmic Pyramid of 400 "steps."

The hidden significance of this 17 "step" gap, in the evolutionary sequence of all Earth-type planets, lies in its ability to resolve a potential paradox with regard to a final merger of "M"-type spirit into a stellar black hole. Without the existence of this gap, which must result in stellar black holes outnumbering any one specific "M" level by a factor of 10^5, we would encounter a somewhat outlandish situation at the more advanced stages of man's evolution. Since theory demands a reduction of numbers in direct proportion to higher status, it follows that the final "M" stages would otherwise be quite literally populated by a mere handful of individuals. The inferred gap of some 17 "steps" will greatly alleviate this problem by assuring that the ultimate "M" stage, immediately prior to achieving stellar black hole fusion, will consist of not less than 10^5 spiritual entities. When combined with the ratio of 10^5 stellar black holes in excess of a solitary "M" stage, we arrive at a total of 10^{10} entities — exactly the sum needed to account for a typical M^1 population level! Hence, it is no wonder that there is a curious similarity in the numbers of stellar black holes and Earth-type planets. It does, in fact, seem to necessitate an apparent "surplus" of 10^5 stellar black holes in order to balance cosmic books in terms of "**A**" content.

To grasp the full picture of this crucial relationship, involving relative abundance and fused "A" content of the various cosmic levels, it is perhaps constructive to assess the total range of "M" stages in some detail. (See Fig. #11.) Immediately prior to our present M^1 stage, it will be noted, the abrupt increase (10^5) in worlds must require a corresponding reduction in the number of pre-human entities. This is precisely what is observed when we examine the distribution of our own terrestrial life forms. The comparative shortage of primitive man is clearly no accident; nor is the recent sharp increase in human population an unprecedented occurrence. In both instances we are likely witnessing nature's balancing of appropriate cosmic equations. (Incidentally, the relative scarcity of early man tends to support the conviction that there is really little difference in spiritual value between the higher animals and primitive man — with the family pet possibly only short steps away on the evolutionary ladder!)

In predicting the future of mankind, as he passes through the successive "M" levels, it is to be inferred that each doubling will involve about 75,000 years in terms of Earth history. After all 17 "M" levels (or generations of human doubling) have been attained, it may be calculated that some 1,275,000 years will have elapsed as man finally concludes his evolutionary saga on the planet Earth. From this point onward our world will continue to feature all present life forms below that of M^1; but any trace of a higher intelligence will likely be conspicuous by its absence. In actuality, *man will have vanished from the face of this planet simply because he succeeded in evolving beyond the human form!* (This deduction would readily explain any difficulty in contacting intelligent life forms on other planets. It is not so much that they are intrinsically scarce in our Galaxy, as it is that large numbers of such civilizations have already come and gone in the relatively short time span allocated for human supremacy.)

Invariably, it is the evolution of the human brain itself which provides the mechanism for his eventual reduction in numbers. Along with the degree of spiritual progress achieved during a past life, the rebirth of a soul must be regulated by two basic factors. On the one hand, we have the quantity of spirit awaiting incarnation; while on the other, we have the number of suitable body-forms available for housing a specific level of spirit. With evolution serving to very gradually change the structure and caliber of man's brain, the incorporation of human souls will become highly restrictive as fewer entities of a particular status seek manifestation in fewer appropriate body-forms. Essentially, nature will be seen to impose its own foolproof method of birth control. As man pursues his ascent of the Cosmic Pyramid, so emphasis will be shifted from quantity to quality. (See Fig. #12.)

Beyond the exalted status of a stellar black hole there may well lie another universe embracing new and unfamiliar laws. Within this lofty domain of *Principle*, spirit is enabled to climax its lengthy evolution by transcending such mundane attributes as gravitation, electricity and time. In the eyes of mortal man the black hole state may very well be worshiped as a wondrous God-like realm of reciprocal communion — from whence Perfection will ultimately ensue as Total Fusion of Black Hole Minds becomes a reality and time ceases to exist! Somehow, as a direct consequence of this momentous transition from imperfection and time, the mass-energy so released is manifested as a new generation of "**A**" within the confines of our physical universe.

Perhaps, of all matter forms subject to instant disintegration at the edge of the universe, the unique properties of black holes could suffice to preserve them from sharing the same abrupt fate. Such an exception might even turn out to be both a logical and a necessary inference — especially in the case of the more massive quasar-type black holes. Ranging in size from several solar masses, for a typical stellar black hole, it is conceivable that a chain of successively larger singularities could prevail up to some billions of times the mass of our Sun. If the supermassive black hole at the center of the giant galaxy M87 (at least several billion solar masses) is any indication, this would seem to represent the upper limit for the size of a black hole. Calculation of the "**A**" content, for an object of this nature, leads to a total of something like 10^{108} fused "**A**"s. For an entity of this evolutionary status its theoretical abundance appears to be much higher than what is observed. This discrepancy is effectively removed, however, by the simple expedient of exempting black holes — at least supermassive quasar-type entities — from any premature extinction at the cosmic periphery.

In any event, a truly amazing picture emerges as to what lies beyond the human form. We are at last in a position to glimpse a future that must exceed our wildest dreams. Upon contemplating the premise of an Eternal Universe, it may well be said that mankind now stands at the threshold of an entirely new and promising field of inquiry, the surface of which has barely been scratched. The challenge to man is clearly without parallel; but the reward will more than justify his effort.

A Cosmic Paradox

While a physical universe of finite size is thus theorized, and with stability assured by reason of an annihilation factor equal to that of creation, a paradox arises when we contemplate a connection with the spiritual aspect of cosmic evolution. If spirit forever rises from an initial state of a lowly "**A**" to one of Perfection (or God), then may it not be said that — with the passage of time — God is becoming increasingly more Omnipotent? Conversely, might it also be argued that God was less of an Entity in the distant past? The issue of a finite physical universe is therefore seen to differ from the issue of spiritual evolution in one crucial respect. In essence, our physical environment may be likened unto that of a catalyst — a medium remaining unchanged in the long run while facilitating an endless progression of spirit up the Great Cosmic Pyramid.

Conceivably, a solution to this Eternity paradox may reside in a property so believed to characterize the Ultimate State of Perfection — namely, a state in which *time* has literally ceased to exist! In the absence of time — where communion is instantaneous — distance and physical dimensions become meaningless terms. Devoid of both time and size, the problem of an ever-increasing and evolving Creator tends to fade into oblivion. Likewise, the property of quantity — being linked to size — is similarly shrouded in nebulosity should there be no such thing as time. Looked at from the point of view of mortal man, there does indeed appear to be a contradiction. At the level of Perfection, human logic and reasoning may very well have no substance.

It is unfortunate that our rather limited faculties prevent us from presenting a more satisfactory answer to this "Cosmic Paradox." However, it is also clear that one who is less than Perfect must be precluded from knowing all. At our current evolutionary level it is, perhaps, sufficient to recognize that any distortion of Reality — through introduction of the phenomenon of time — is capable of posing many seeming paradoxes. One can only speculate as to when and at what stage of our evolutionary pilgrimage this particular mystery will yield its secret.

Summary

It is certainly quite remarkable that galaxy associations, with highly discordant redshifts, can be made to give very plausible separations in terms of principles inherent in the "New Cosmology." And yet, this is but one example of a host of celestial anomalies which may be resolved upon the basis of a revised concept of radiation propagation and the premise of a gravitational repulsion factor. In particular, we now have before us an explanation for our expanding universe without recourse to a Big Bang cosmology.

Once considered an adequate explanation for the phenomenon of gravitation, Newtonian mechanics is now seen to require a degree of modification by no less than three additional equations. By way of recapitulation, it may be stated that this enigmatic force of nature is likely to be conducive to expression as follows:

(1) The traditional Newtonian *inverse-square* law.
(2) A so-called relativistic *adjustment for motion.* (In reality, this formula depicts the declining efficiency of gravitation with an increase in relative motion.)
(3) The prescribed *distance/repulsion* (D/R) formula, in which

a portion of exchanged energy is deemed to repel.

(4) A *time-lapse* factor (for which exact calibration is pending), as no physical reaction can take place in zero time.

Noting that the inferred D/R factor can account for the expansion of the universe — without recourse to a Big Bang cosmology — it is now evident that this same premise is capable of clarifying a variety of "missing mass" mysteries. In turn, a study of quasar distribution — in the light of revised views of radiation propagation — has served to reveal a startling picture of a fateful cosmic edge linked closely to the issue of creation and reincarnation.

Science seems to have a way of progressing in cycles, with periods of stagnation interspersed with sudden leaps forward in response to new ideas and advanced technology. There is every reason to believe that conditions are favorable to remedy certain misconceptions that have clouded such fundamental issues as gravitation, radiation propagation and cosmic origin. Hopefully, in assessing all of the many revolutionary implications of the "New Cosmology," the scientific community will not allow the instinctive prejudice of man to refute ideas simply because they require admission of past error.

But a truly monumental and totally unexpected challenge which now confronts mankind is for astronomers to more accurately determine our proximity to an impenetrable shell of trapped energy so believed to enclose our spherical universe — a barrier which threatens to one day extinguish the planet Earth with the swiftness of a lightning bolt! Nor may it be thought that the gauntlet has been cast only to members of this noble profession. In the years to come, there are crucial roles to be played by both philosophers and physicists alike. Indeed, it may be inferred that the evolving science of cosmology, while currently in its infancy, is destined to become more and more the realm of philosophers.

Not the least intriguing of the challenges confronting future generations of philosophers is that of refining details relating to the universal "time constant." Very conceivably, in pursuing such concepts as reincarnation and cosmic fusion, the modern philosopher could become the high priest of scientific research.

News Note:

In an informative study of gamma-ray bursts (GRBs), it is revealed (*Sky & Telescope*, August, 2006) that the more powerful bursts are also those of higher redshift. On the surface, this would seem to support a Big Bang cosmology in which luminosity is interpreted as a function of age, since it can be shown that it is the metal-poor younger stars which produce the more luminous GRBs. However, as with detection of "surplus" distant radio sources, it is really a case of observing more than one generation of creation— hence, more of the massive metal-poor stars giving rise to powerful GRBs.

Appendix 1.

Extragalactic Hydrogen Clouds

By reason of their intense luminosity, which is often a hundred times or more than that of a giant galaxy, quasars can be readily seen at great distances. En route to us across the depths of space, this radiation is bound to interact with vast clouds of hydrogen gas — of comparatively recent vintage — which are mandated by the Steady-state model of creation. As a quasar's light passes through such a cloud, the hydrogen will absorb light at very specific wavelengths, with the most notable of these discrete absorptions being the Lyman-alpha line located at a rest wavelength of 1,216 angstroms. But since all extragalactic hydrogen clouds share in the expansion of the universe, it follows that such absorption lines will be shifted to longer wavelengths commensurate with the velocity of recession and distance of these intervening clouds.

Upon analysis of Lyman-alpha lines in the spectra of a sampling of many distant quasars, it was found that the number of hydrogen clouds increases (with higher redshift) at a rate greater than expected from a simple increase of volume with distance. As was the case with an apparent "surplus" of radio sources, this has been interpreted as evidence supporting Big Bang cosmology, since it is deemed that these clouds were far more numerous in the remote past when the universe was supposedly only a fraction of its present size and much of the hydrogen had yet to condense into stars and galaxies. Although now seen to have a perfectly logical explanation upon the basis of the "New Cosmology," it becomes of paramount importance to examine Lyman-alpha absorption lines at relatively low redshift. (Should an appreciable number of uncondensed hydrogen clouds be found at modest redshifts, it would only be explicable in terms of the Steady-state principle featuring newly created matter.)

Prior to the Hubble Space Telescope, it was not possible for astronomers to observe Lyman-alpha lines much below Z=2.00, as such radiation is absorbed by the Earth's atmosphere. However, the results are now available and reveal the presence of hydrogen gas clouds inhabiting the regions between nearby superclusters. By way of an example, the ultraviolet spectrum of 3C273 (about 2 BLY away) reveals no less than five major intervening hydrogen clouds. Moreover, their spacing is remarkably uniform and is in excellent agreement with separations to be inferred from a Steady-state/Small Bang creation model! [1]

Clearly, evidence that was first thought to favor the Big Bang concept has once more been disclosed to lend strong support on behalf of the principle of Continuous Creation.

Reference:

[1] "A Quasar Lights up the Universe," *Astronomy*, September 1991, pps. 42-45.

Appendix 2.

A "Cosmological Constant"

Impressive evidence against Big Bang cosmology is forthcoming from a study of type 1a supernovae. This type of supernova is produced when a white dwarf star acquires a critical amount of gas from an evolving binary companion. When the mass of the degenerate dwarf star reaches some 1.4 times that of our Sun, it will literally explode with predictable violence. Since the factor of luminosity is a function of mass, such explosions are of remarkably similar brilliance. As a result, this allows astronomers to use these objects as a reliable means of determining distance quite independent of redshift.

To the astonishment and consternation of many scientists, it has now become necessary to postulate a *negative* side to gravitation in order to account for observation. (The highly respected journal, *Science*, cites this discovery of a "cosmological constant" as "the most important scientific advance of 1998.") What has caused this sudden conundrum among certain members of the scientific community? As reported in a number of astronomy journals, studies of distant type 1a supernovae have revealed that expansion of the universe is not slowing down with the passage of time. (This slowing of cosmic expansion is a firm requirement of traditional viewpoint, which would consider gravitation to be solely an attractive force propagated indefinitely into space.) Instead, supernovae research has disclosed that the universe is expanding at a rate consistent with the premise that if you double the distance you also double the speed of recession. In turn, this must indicate a steady repulsive factor and rule out any Big Bang cosmology, leaving us with only one alternative — an Eternal or Steady-state universe featuring the principle of renewal through the process of Continuous Creation.

It is somewhat ironic that the very concept of a "cosmological constant" (or D/R factor), which Einstein later considered to be his "greatest mistake," should now be given the credence which it has always deserved. Hopefully, such observation will serve to bring cosmologists back to reality and to finally bury, once and for all, the idea of a Big Bang creation.

Incidentally, the premise of a D/R factor leads to yet another prediction capable of verification through observation. By reason of a steady outflowing of quanta from the vicinity of the cosmic center, there is bound to be a tendency for newly forming galaxies to adopt a common orientation of their axis of rotation— namely, one whereby a galaxy's polar axis will tend to be aligned in a center/edge direction.

This preferred orientation may be ascribed to the way in which the flow of D/R energy impacts a roughly spherical cloud of condensing gas. Should cloud rotation become elongated in a contrary direction, D/R pressure will act to impose repulsion in a conflicting manner to opposite edges of such elongated systems. However, if the axis of rotation was to line up with the cosmic center/edge, there would be no opposing influence. Hence, it is to be inferred that a preponderance of elliptical and spiral galaxies should exhibit traces of this common axis of rotation— in spite of any later influence imposed by interaction with companion systems. (This prediction has since been verified by observation. See the *Astrophysical Journal Letters*, April 1, 2006.)

Appendix 3.

The "Principle of Equivalence" Law

An interesting observation has come to light with regard to NASA's Pioneer space probes. (See "A Space Mystery," *Newsweek*, October 4, 1999.) Contrary to expectation, the two probes (moving outward in nearly opposite directions) are slowing down at a rate which cannot be explained upon the basis of current laws of gravitation.

Although this discrepancy would appear to be only one part in ten billion, at the distances of the twin Pioneer probes the actual modification of gravity needed is likely of the order of one part in 170,000. The additional ingredient may be ascribed to the inferred distance/repulsion (D/R) aspect of gravitation. The big question which now arises is why the probes are slowing down instead of accelerating, as might be expected. To grasp why this is so we must consider the *cumulative* effects of what may be termed a fourth expression of gravitation — namely, a *time lapse* factor in which it is conceded that no physical reaction can take place in zero time. Since the force of gravitation is believed to propagate — at close to the speed of light — from atom-to-atom, it follows that a massive body will resist acceleration more than a tiny object. Invariably, the "principle of equivalence" law, long cherished by physicists, will have to be modified accordingly. (This effect is discussed in Chapter 5, under the caption of "The Microwave Background.")

Appendix 4.

Calibration of Hubble's Constant

Efforts to calibrate Hubble's constant (the rate by which the universe is expanding) have been at the frontiers of astronomical research for well over half a century. Today, the consensus of opinion among most members of the astronomical community, is that the universe has an age/radius of somewhere between 10 and 20 billion years. However, this lower figure is in conflict with evidence that the oldest stars of our Milky Way's globular clusters were formed at least 15 billion years ago. For this reason, an antiquity well beyond 10 billion years is to be preferred by Big Bang proponents in an attempt to reconcile cosmic age with the firm dictates of stellar evolution.

As it turns out, there is conclusive proof of Big Bang fallacy merely by pausing to examine redshifts within the confines of our own Local Supercluster. It so happens that the average redshift (per MLY of separation) is of the order of 22 km/s. But since members of our supercluster are constrained by gravitational attraction, *this would require the true rate of cosmic expansion to be a great deal higher!* Therein lies an insoluble problem for supporters of all forms of Big Bang cosmology, since admission of higher redshifts (per MLY) only serves to compound an already serious contradiction. (A redshift of 17 km/s would imply an age/radius of about 18 BLY; while 22 km/s gives a figure of 14 BLY; 30 km/s implies 10 BLY; and 40 km/s only 7.5 BLY.)

The problem for Big Bang proponents gets rapidly worse when we assess galactic redshifts in relation to distance traveled since the time of their birth. Should we tentatively adopt a figure of 15 BY as being consistent with both Hubble's constant and the ages of our oldest stars, then it may be calculated that a system now 20 MLY distant (with an average velocity of recession of some 330 km/s over the past 15 BY) will have receded about 16.5 MLY since forming out of the interstellar medium. A galaxy currently situated 40 MLY distant will have moved away by not less than 33 MLY; while another at 100 MLY will be adduced to have traveled over 82 MLY in this very same interval. Should we go back in time to their formation we are faced with the rather ludicrous situation of having condensed our entire Local Supercluster, consisting of thousands of galaxies, into a volume of space no more than 30 to 40 MLY in diameter! Such enormous density would clearly possess a gravitational field so strong that no system could ever manage to escape to their observed distances. Upon the basis of a solitary Big Bang explosion there is simply no way in which they could have receded to their present positions — revealing that there is something drastically wrong with the idea that such galactic redshifts depict true motion of recession. This radiation does, in fact, reflect a substantial *time delay* factor imposed by reason of interaction with the inferred D/R quanta and the ubiquitous interstellar medium.

However, while conclusively ruling out the Big Bang model, upon the premise of a Small Bang/Steady-state cosmology, it is logical to expect that a portion of these redshifts will actually be due to residual movement away from the site of our Local Supercluster Small Bang explosion — an event which is presumed to have occurred over 15 billion years ago in the vicinity of M87, at the heart of the dense Virgo Cluster. Clearly, the task before us is to assess the relative importance of these two factors. Although much study will be needed in order to more fully resolve this issue, it is conceivable that an estimate of 50% actual motion (away from M87) and an equal proportion of "time delay" may not be entirely out of line. (If so, it is likely that velocities of roughly 1,000 km/s will have to be subtracted from the redshifts of distant celestial objects to allow for passage of radiation through our own supercluster, as it is met by both outflowing D/R quanta and the interstellar medium.)

Thus it will be seen that galactic redshifts cannot be used as a means of establishing Hubble's constant through direct extrapolation. Only by measuring the redshifts of superclusters — over distances of several billion light-years — may we make a proper determination of this vital cosmic constant. Even then, it will be necessary to consider this "time delay" component, inherent in all extragalactic redshifts, before such measurements can be deemed truly trustworthy.

Meanwhile, based upon available criteria, it seems reasonable to presume that a figure of roughly 17 km/s (per MLY) may be adopted as being as good as any other in mapping our universe. This would yield a cosmic radius of some 18 BLY — being the point where recession is equal to the speed of light relative to the center of the universe. Hubble's constant, as described previously, would then translate to some 12 billion years.

Fig, 1. Division of Geological Time

PERIOD & EPOCH		LIFE FORMS	YEARS AGO (millions)
Quarternary	Recent	Primitive Man	2+
	Pleistocene		12
		Early Primates	26
	Pliocene		
	Miocene	Mammals	
Tertiary	Oligocene	have replaced	
	Eocene	Dinosaurs	
	Paleocene		65
Cretaceous		Flowering	136
		Plants	
Jurassic		Primitive Birds	195
		and Mammals	
Triassic		Dinosaurs	225
Permian		Reptiles	280
		Insects	320
Carboniferous		Lush Forests	350
		Land Ferns	
Devonian		Amphibians	400
		Many	
Silurian		Fishes	430
Ordovician		Trilobites	500
		Variety of	
Cambrian		Marine Life	570
		Invertebrates	700
		Simple Cells	
Precambrian Era		Elementary	3,500+
		Life Forms	

***Fig, 2.* Redshift/Time/Velocity/Distance Scale**
(Concentric View)

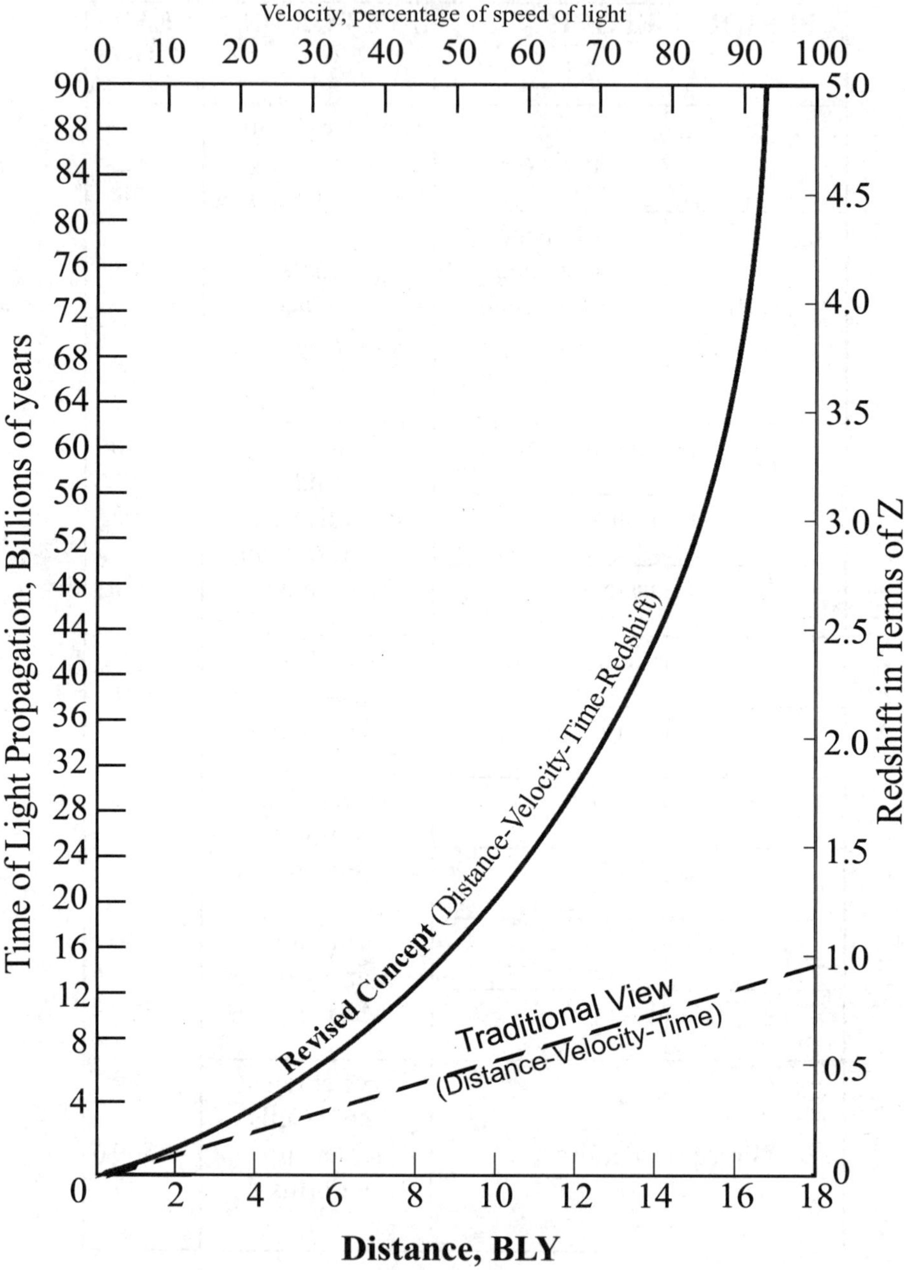

Fig, 3. **Expansion of Universe and Propagation of Light**

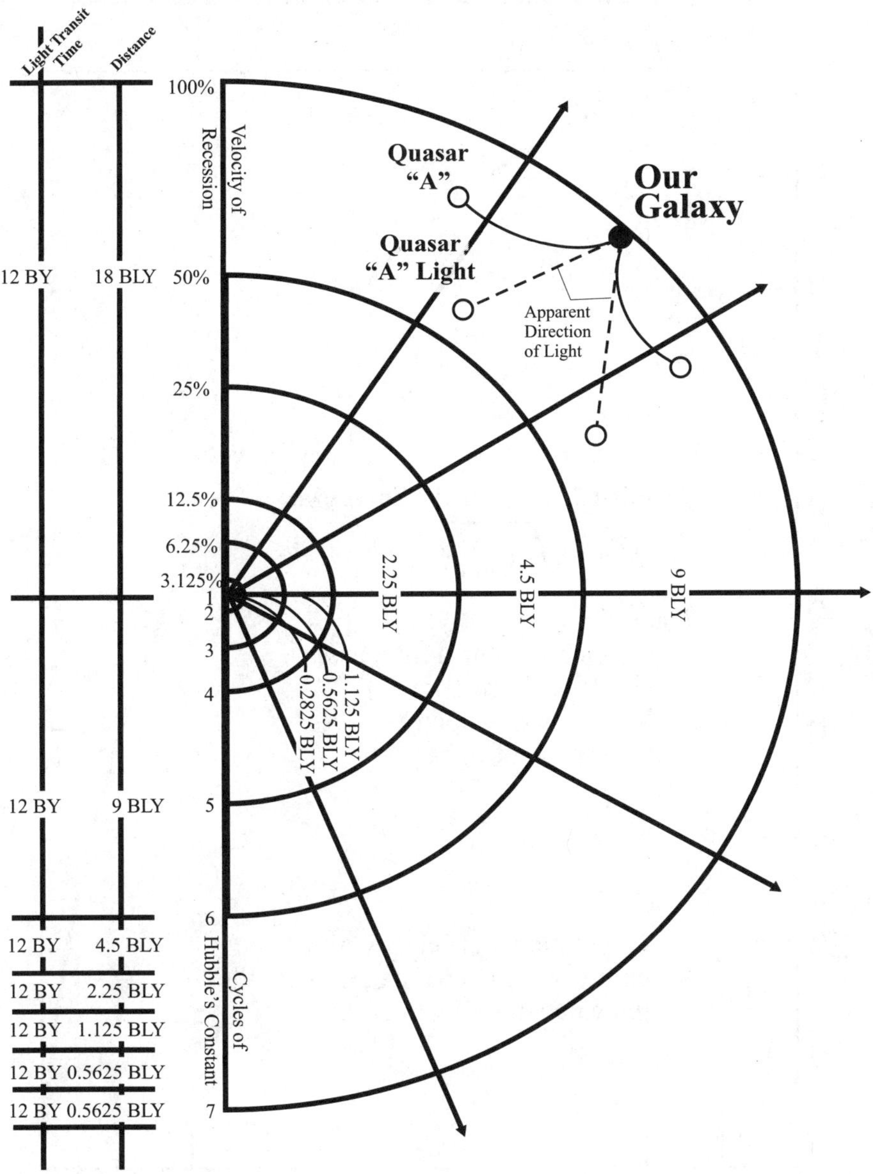

Fig, 4. Efficiency of Gravitation with Relative Motion

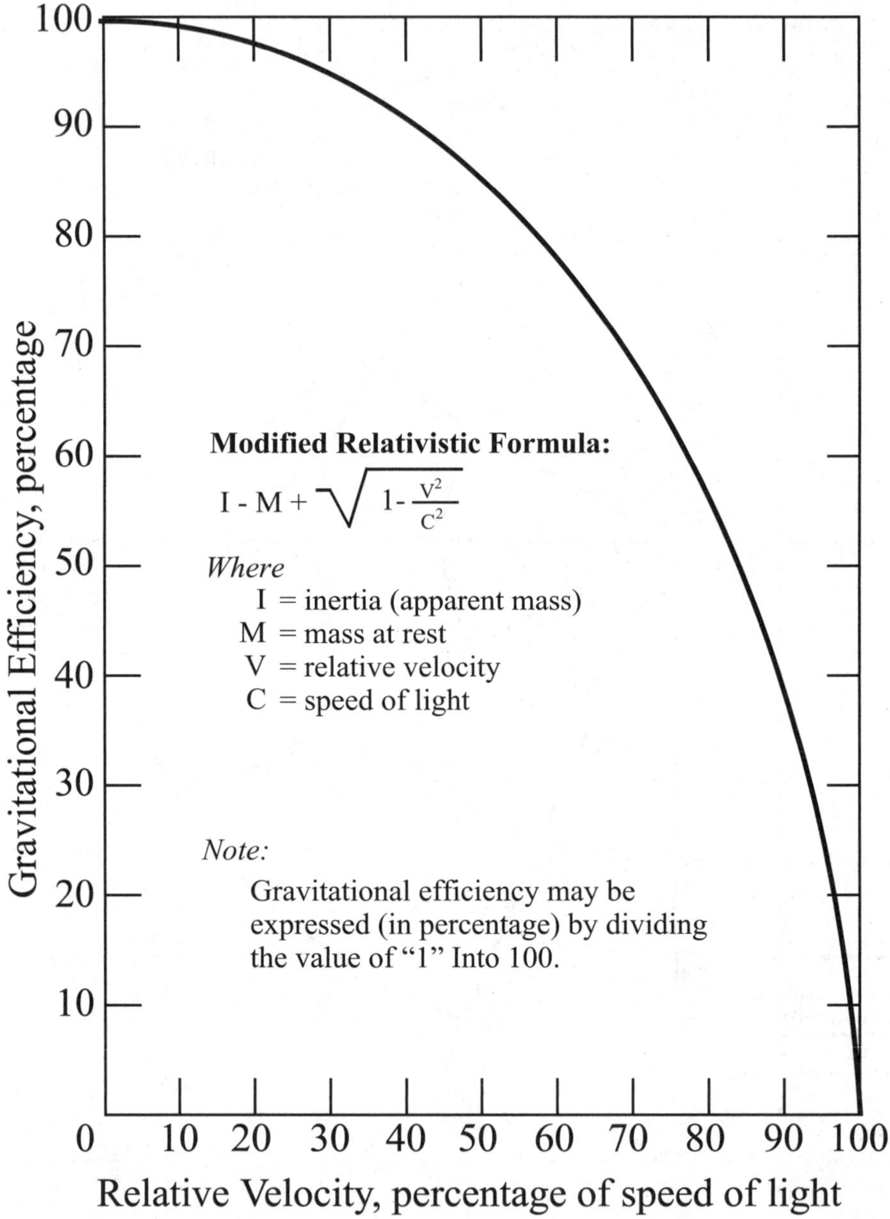

Fig. 5. Distance/Repulsion Relationship of Gravitation

(Distance) Fourfold Increase (from proton nucleus)	**(Repulsion) Twofold Increase** (in terms of attraction)	**(Distance) Fourfold Increase** (from proton nucleus)	**(Repulsion) Twofold Increase** (in terms of attraction)
1.5×10^{-13} cm	10^{-20}	110.68 km	8.590×10^{-10}
6.0×10^{-13} cm	2.0×10^{-19}	442.72 km	1.718×10^{-9}
2.4×10^{-12} cm	4.0×10^{-19}	1,770.89 km	3.436×10^{-9}
9.6×10^{-12} cm	8.0×10^{-19}	7,083.55 km	6.872×10^{-9}
3.84×10^{-11} cm	1.6×10^{-18}	28,334 km	1.374×10^{-8}
1.536×10^{-10} cm	3.2×10^{-18}	113,337 km	2.749×10^{-8}
6.144×10^{-10} cm	6.4×10^{-18}	453,347 km	5.498×10^{-8}
2.458×10^{-9} cm	1.28×10^{-17}	1.813×10^{6} km	1.100×10^{-7}
9.830×10^{-9} cm	2.56×10^{-17}	7.254×10^{6} km	2.200×10^{-7}
3.932×10^{-8} cm	5.12×10^{-17}	2.901×10^{7} km	4.398×10^{-7}
1.573×10^{-7} cm	1.024×10^{-16}	1.161×10^{8} km	8.796×10^{-7}
6.291×10^{-7} cm	2.084×10^{-16}	4.642×10^{8} km	1.759×10^{-6}
2.517×10^{-6} cm	4.096×10^{-16}	1.857×10^{9} km	3.518×10^{-6}
1.007×10^{-5} cm	8.192×10^{-16}	7.428×10^{9} km	7.037×10^{-6}
4.027×10^{-5} cm	1.638×10^{-15}	2.971×10^{10} km	1.407×10^{-5}
1.611×10^{-4} cm	3.277×10^{-15}	1.188×10^{11} km	2.815×10^{-5}
6.442×10^{-4} cm	6.554×10^{-15}	4.754×10^{11} km	5.629×10^{-5}
2.577×10^{-4} cm	1.311×10^{-14}	1.901×10^{12} km	1.126×10^{-4}
.010308 cm	2.621×10^{-14}	7.606×10^{12} km	2.252×10^{-4}
.041232 cm	5.243×10^{-14}	3.2158 LY	4.504×10^{-4}
.164927 cm	1.049×10^{-13}	12.8634 LY	9.007×10^{-4}
.659707 cm	2.097×10^{-13}	51.4535 LY	1.801×10^{-3}
2.63883 cm	4.194×10^{-13}	205.814 LY	3.603×10^{-3}
10.5553 cm	8.389×10^{-13}	823.257 LY	7.206×10^{-3}
42.2212 cm	1.678×10^{-12}	3,293,03 LY	.01441
168.885 cm	3.355×10^{-12}	13,172.1 LY	.02882
675.540 cm	6.711×10^{-12}	52,688.4 LY	.05765
2,702.2 cm	1.342×10^{-11}	210,754 LY	.11529
10,809 cm	2.684×10^{-11}	843,015 LY	:23058
43,235 cm	5.369×10^{-11}	3,372,060 LY	.46116
1.72938 km	1.074×10^{-10}	13,488,239 LY	.92234
6.91763 km	2.147×10^{-10}	(16,900,000 LY)	(1.0000)
27.6701 km	4.295×10^{-10}		

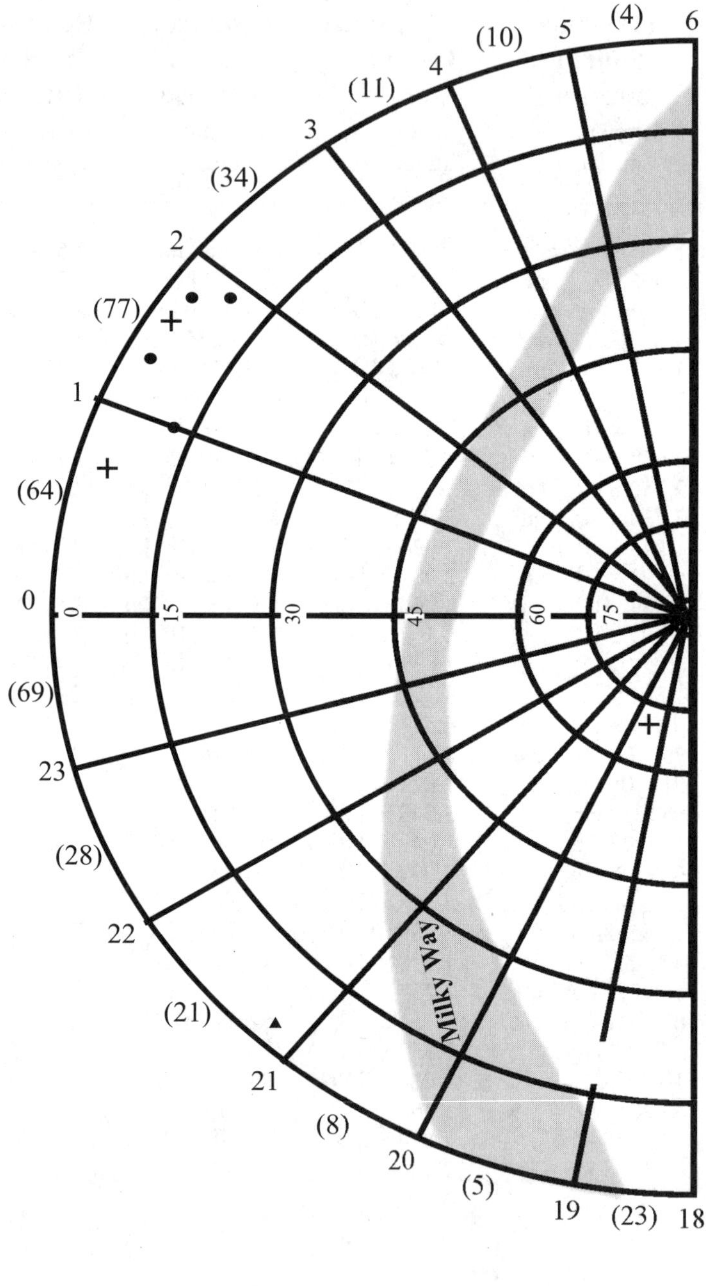

Fig, 6. Northern Hemisphere Quasars

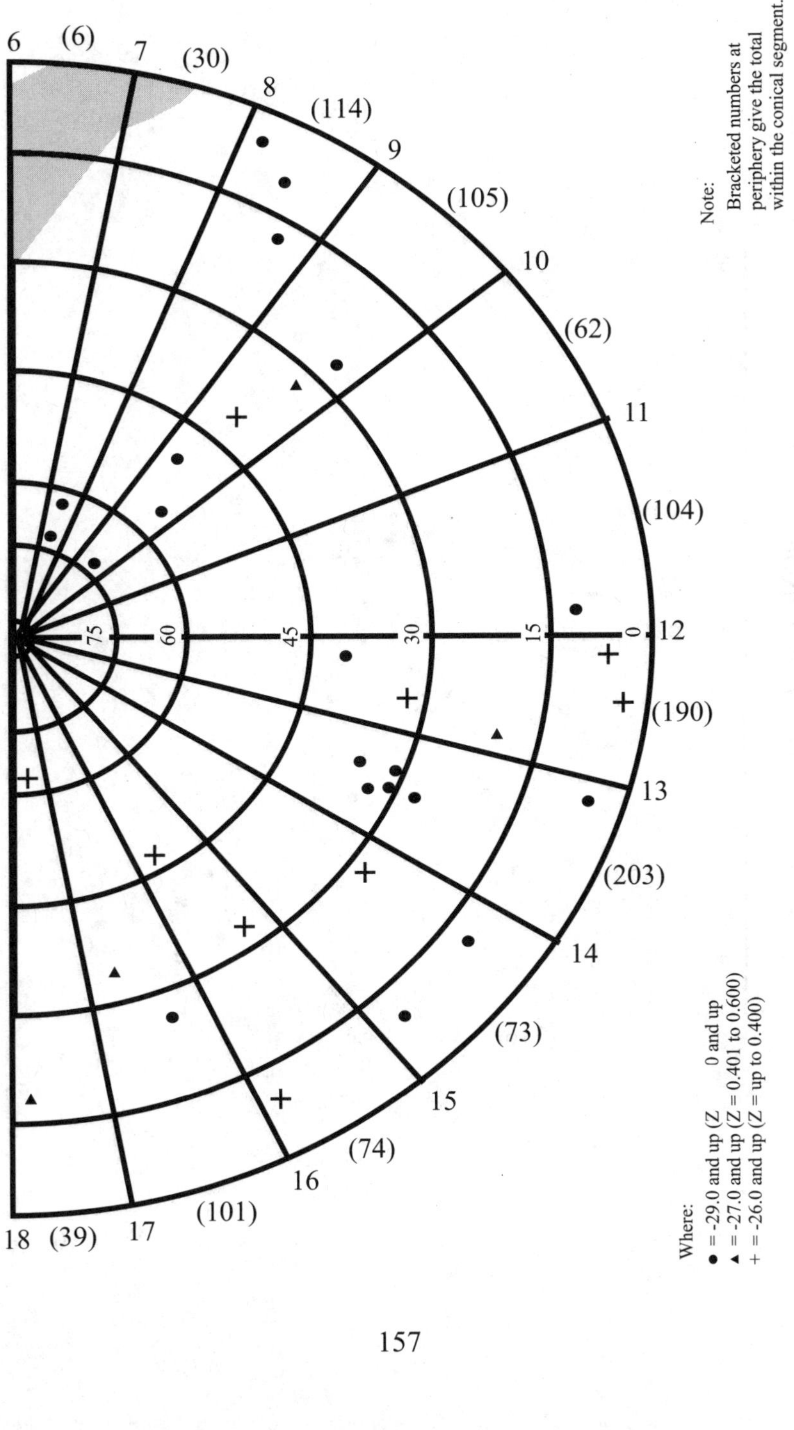
6
(6)
7
(30)
8
(114)
9
(105)
10
(62)
11
(104)
12
(190)
13
(203)
14
(73)
15
(74)
16
(101)
17
(39)
18
75
60
45
30
15
0
Note:
Bracketed numbers at periphery give the total within the conical segment.
Where:
● = -29.0 and up (Z 0 and up
▲ = -27.0 and up (Z = 0.401 to 0.600)
+ = -26.0 and up (Z = up to 0.400)

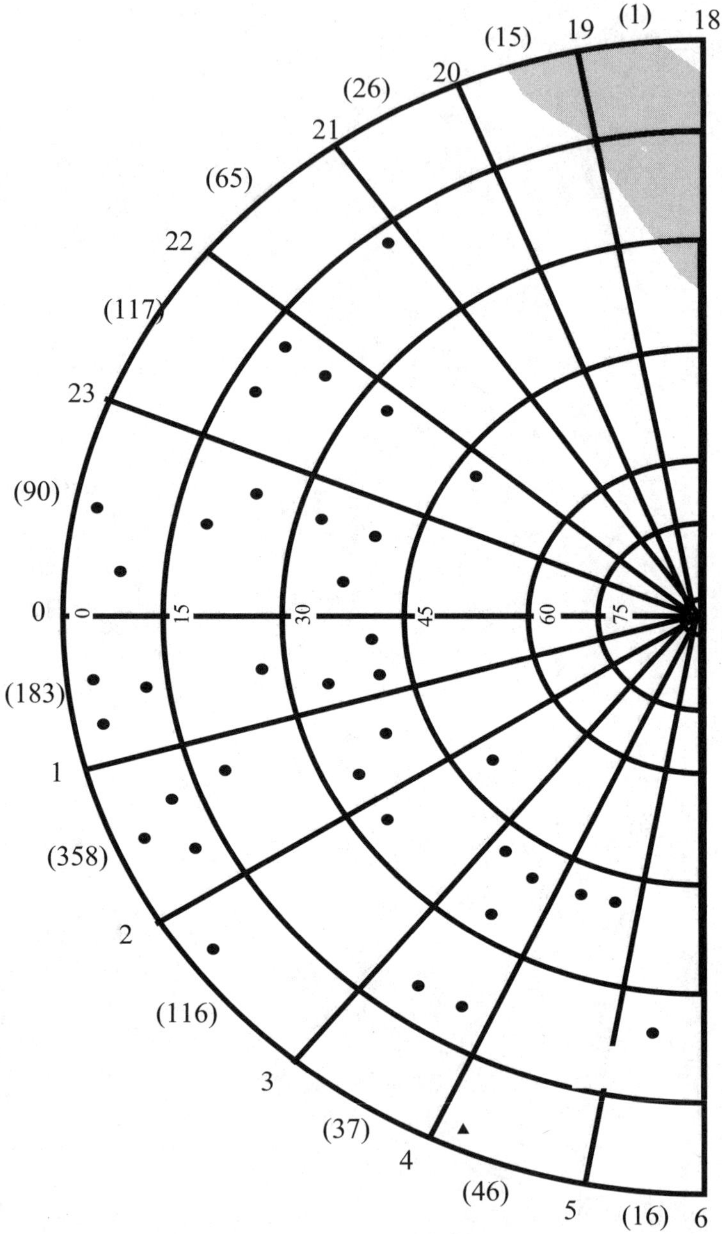

Fig, 7. Southern Hemisphere Quasars

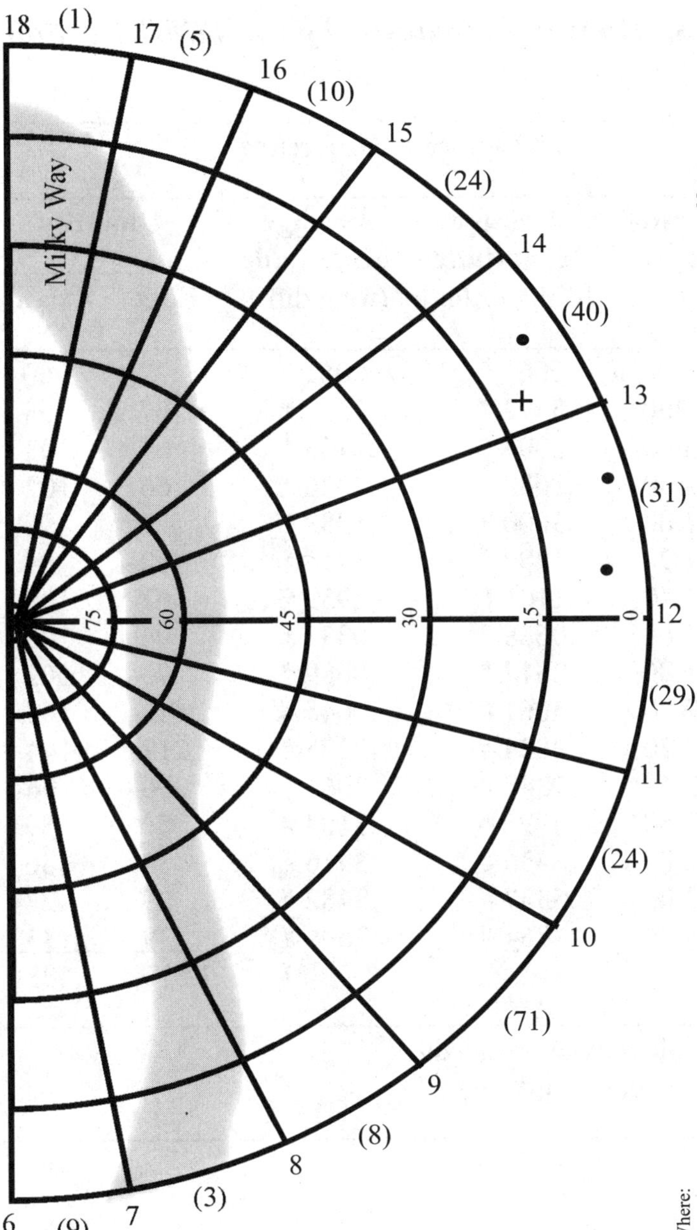

Note:
Bracketed numbers at periphery give the total within the conical segment.

Where:
● = -29.0 and up (Z   0 and up
▲ = -27.0 and up (Z = 0.401 to 0.600)
+ = -26.0 and up (Z = up to 0.400)

Fig. 8. Quasar Luminosity by Redshift/Direction

Magnitude Difference				
Redshift Range Z=	**Average Magnitude** (diff. per bin)	**Average Magnitude** (total diff.)	**Quantity**	
			Edge	**Balance**
up to .200	.0061 *	.0061 *	22	60
.200 to .400	.5572 *	.3232 *	67	98
.400 to .600	.2007 *	.2458 *	67	103
.600 to .800	.1688 *	.2076 *	66	102
.800 to 1.00	.3000 *	.1788 *	71	90
1.00 to 1.20	.1200 *	.1715 *	60	94
1.20 to 1.40	.1101 *	.1959 *	61	109
1.40 to 1.60	.0688 *	.0887 *	96	100
1.60 to 1.80	.2517 *	.0689 *	91	102
1.80 to 2.00	.3061 #	.0645 #	140	145
2.00 to 2.20	.2824 #	.2275 #	212	144
2.20 to 2.40	.1348 #	.3061 #	206	106
2.40 to 2.60	.4996 #	.3108 #	50	59
2.60 to 2.80	.8436 #	.3316 #	38	36
2.80 to 3.00	.5337 #	.3482 #	29	22
over 3.00	.8764 #	.3696 #	39	35
			1,315	1,405

* = Brighter away from edge.
# = Brighter toward edge.

Fig, 9. Redshift/Time/Velocity/Distance Scale (Eccentric View)

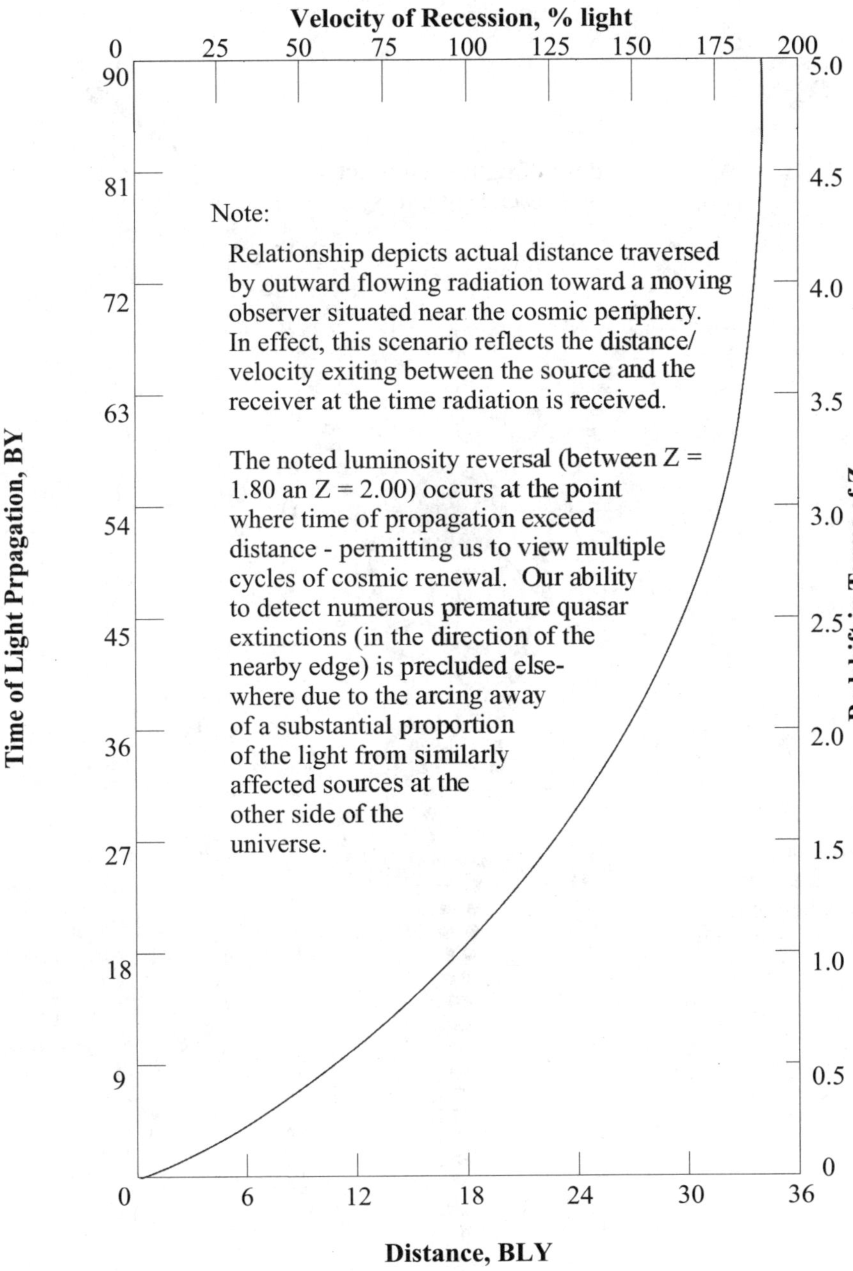

Fig, 10. Galactic Arcing of Radiation

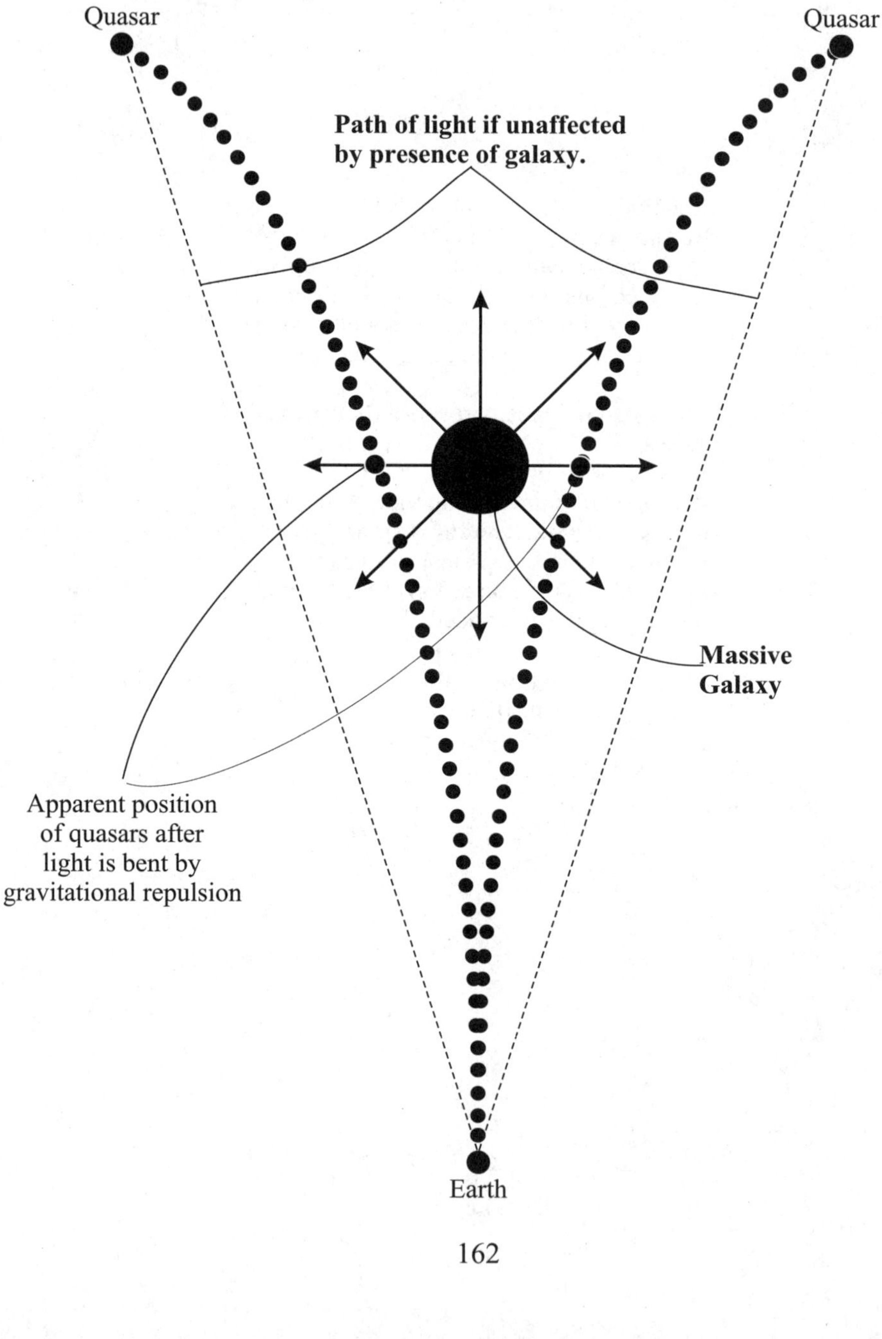

Figure #11

Evolution of "M" Stages

Status or "Steps"	Status "M" Stages	Quantity of Entities	Entity "A" Content	Number of Worlds	Quantity per World
309	Pre-man	$2x10^{27}$	$5x10^{92}$	10^{22}	$2x10^{5}$
310	M^{1}	10^{27}	10^{93}	10^{17}	10^{10}
311	M^{2}	$5x10^{26}$	$2x10^{93}$	10^{17}	$5x10^{9}$
312	M^{3}	$2.5x10^{26}$	$4x10^{93}$	10^{17}	$2.5x10^{9}$
313	M^{4}	$1.25x10^{26}$	$8x10^{93}$	10^{17}	$1.25x10^{9}$
314	M^{5}	$6.25x10^{25}$	$1.6x10^{94}$	10^{17}	$6.25x10^{8}$
315	M^{6}	$3.12x10^{25}$	$3.2x10^{94}$	10^{17}	$3.12x10^{8}$
316	M^{7}	$1.56x10^{25}$	$6.4x10^{94}$	10^{17}	$1.56x10^{8}$
317	M^{8}	$7.81x10^{24}$	$1.28x10^{95}$	10^{17}	$7.81x10^{7}$
318	M^{9}	$3.91x10^{24}$	$2.56x10^{95}$	10^{17}	$3.91x10^{7}$
319	M^{10}	$1.95x10^{24}$	$5.12x10^{95}$	10^{17}	$1,95x10^{7}$
320	M^{11}	$9.77x10^{23}$	$1.02x10^{96}$	10^{17}	$9.77x10^{6}$
321	M^{12}	$4.88x10^{23}$	$2.05x10^{96}$	10^{17}	$4.88x10^{6}$
322	M^{13}	$2.44x10^{23}$	$4.10x10^{96}$	10^{17}	$2.44x10^{6}$
323	M^{14}	$1.22x10^{23}$	$8.19x10^{96}$	10^{17}	$1.22x10^{6}$
324	M^{15}	$6.10x10^{22}$	$1.64x10^{97}$	10^{17}	$6.10x10^{5}$
325	M^{16}	$3.05x10^{22}$	$3.28x10^{97}$	10^{17}	$3.05x10^{5}$
326	M^{17}	$1.53x10^{22}$	$6.55x10^{97}$	10^{17}	$1.53x10^{5}$
327	Stellar Black Hole	$7.63x10^{21}$	$1.31x10^{98}$	10^{22}	1

Fig, 12. Pyramid Structure of the Universe

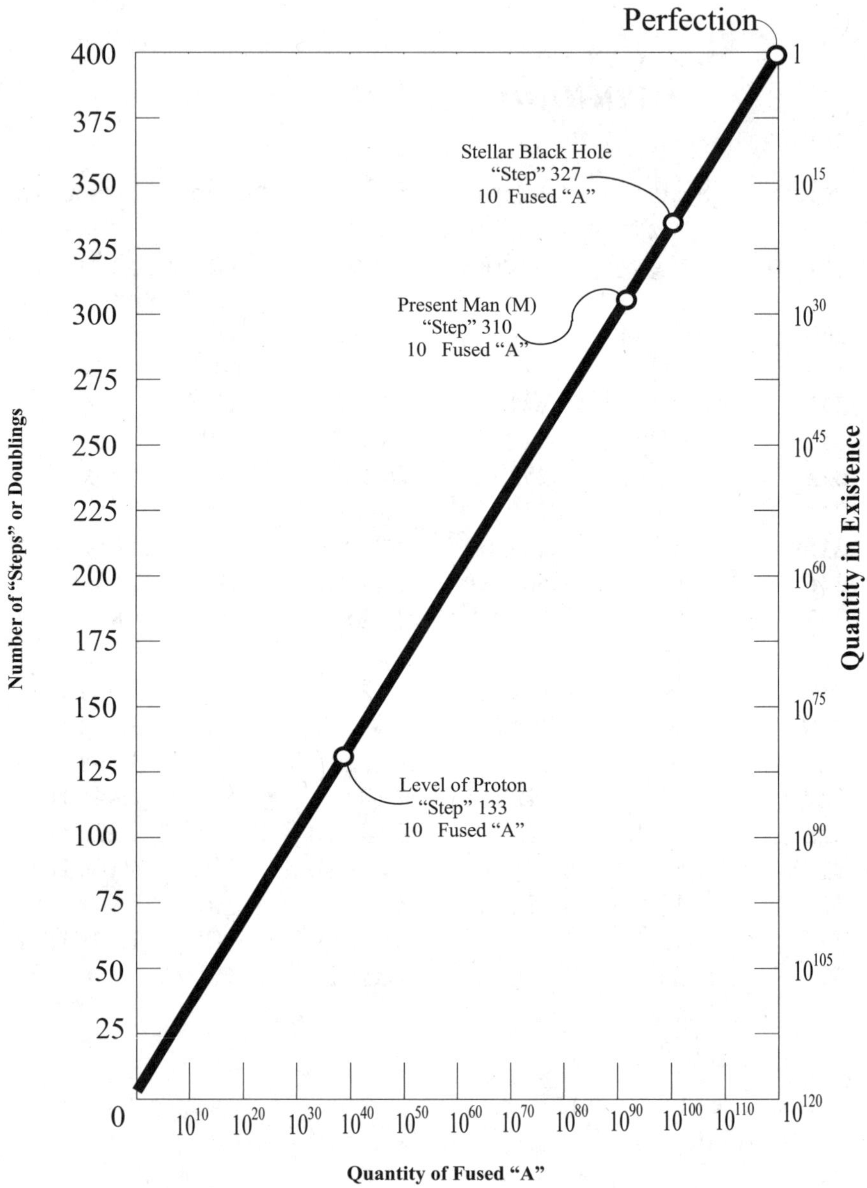

Table 1. Average Galaxy Redshift (per MLY) in Terms of Distance/Direction

Group	Under 50 MLY	50 to 100 MLY	Over 100 MLY
Edge Group:			
0,1,2,23 hr	25.25 km/s	25.10 km/s	24.69 km/s
-0 to -80 deg	(36)	(101)	(20)
Average Group Redshift (all 157 galaxies) = 25.02 km/s			
Adjacent Edge:			
3,4,21,22 hr	23.48 km/s	24.36 km/s	24.17 km/s
-15 to -65 deg	(69)	(135)	(35)
Average Group Redshift (all 239 galaxies) = 24.16 km/s			
Center Group:			
11,12,13,14 hr	26.81 km/s	17.38 km/s	20.00 km/s
+0 to +80 deg	(153)	(405)	(140)
Average Group Redshift (all 698 galaxies) = 19.09 km/s			
Adjacent Center:			
9,10,15,16 hr	20.02 km/s	17.94 km/s	20.61 km/s
+15 to +65 deg	(27)	(116)	(47)
Average Group Redshift (all 190 galaxies) = 19.89 km/s			
Right Angle:			
6,7,18,19 hr	21.17 km/s	20.51 km/s	20.93 km/s
+0 to +80 deg	(107)	(103)	(104)
-0 to -80 deg			
0,1,2,23 hr	*Average Group Redshift (all 314 galaxies)*		
+25 to +65 deg	*= 20.81 km/s*		
11,12,13,14 hr	Note:		
-25 to -65 deg	Quantity within each bin is shown in brackets		

Table 2. Average "Center Group" Galaxy Redshift (per MLY)

Distance (MLY)	Quantity	Average Redshift (km/s)
0 to 30	92	25.30
30 to 50	61	27.91
50 to 60 (Virgo Cluster)	226	18.17
60 to 80	92	15.30
80 to 100	87	17.93
Over 100	140	20.00

With the Virgo Cluster strongly concentrated within the 50 to 60 MLY bin, an informative picture emerges when we examine the adjoining bins which extend some 20 MLY on either side. Were gravitational attraction the sole factor responsible for the noted redshift differential, then why is the closer of the two neighboring bins accelerated some 3.4 times that of the other? Clearly, gravitation is not the only ingredient, and a *non-velocity* factor must be acknowledged. (It is contended that this missing influence is due to the interaction of radiation with D/R quanta. While a favorable flow can assist the propagation of such corpuscles, encounters with quanta ejected from our own Milky Way serve to retard approaching extragalactic radiation — thereby imparting a redshift which, in turn, may be defined as a measure of extended transit time.)

Revelations

Discordant Redshift Association

Spiral galaxy **NGC 450** (redshift 1,863 kn/s), and **UGC 807** (redshift 11,587 kn/s), a small elliptical system. (Photo courtesy of Halton C. Arp.)

Discordant Redshift Association

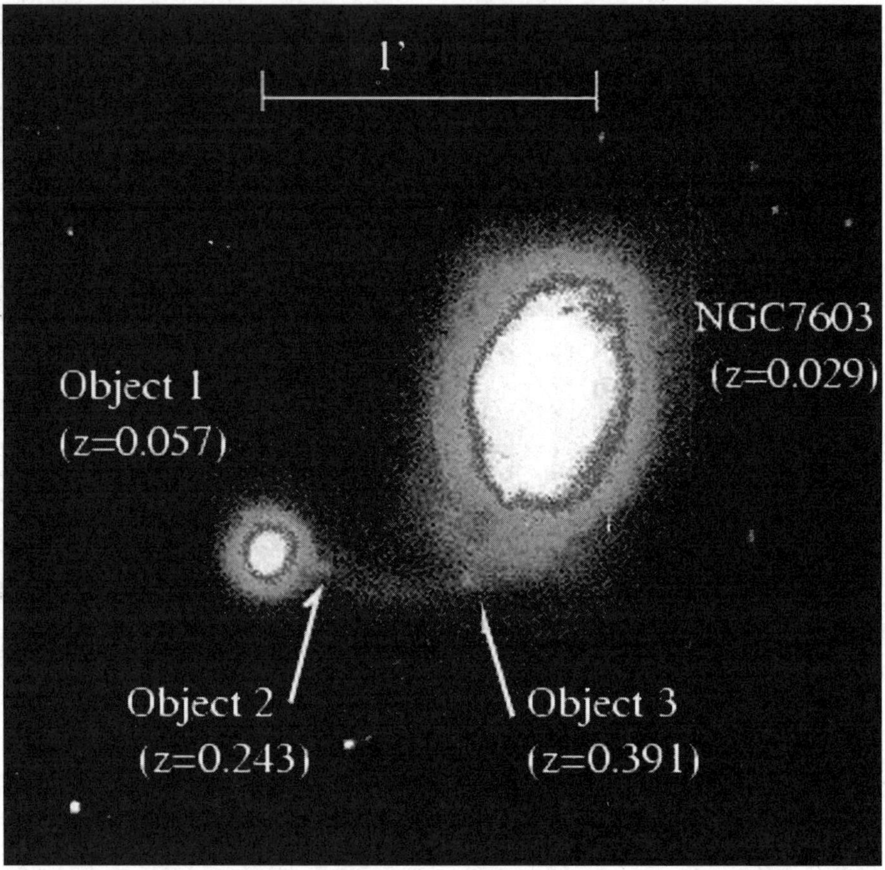

NGC 7603 (redeshift Z=.029) with a small companion system (redshift Z=.057). Embedded in connecting band of stars and gas are Objects #2 and #3, with respective redshifts of Z=.243 and Z=.391. (Photo courtesy of Halton C. Arp.)

Discordant Redshift Association

Spiral galaxy **NGC1232** (redshift 1,776 km/s) with a small companion system (redshift 6,552 km/s)situated at bottom left arm. (Photo courtesy of Halton C. Arp.)

Revelations

Discordant Redshift Association

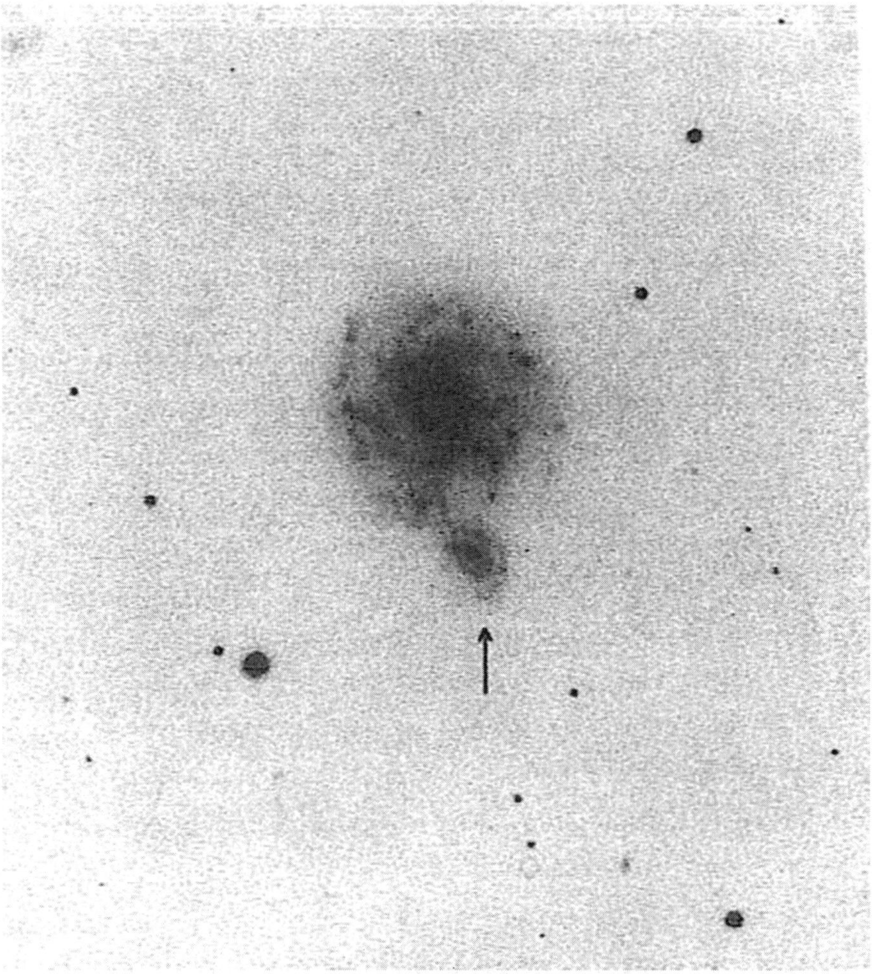

Spiral galaxy **AM 0328-222** and small companion system with a redshift differential of some 17,925 km/s. (Photo courtesy of Halton C. Arp)

Discordant Redshift Association

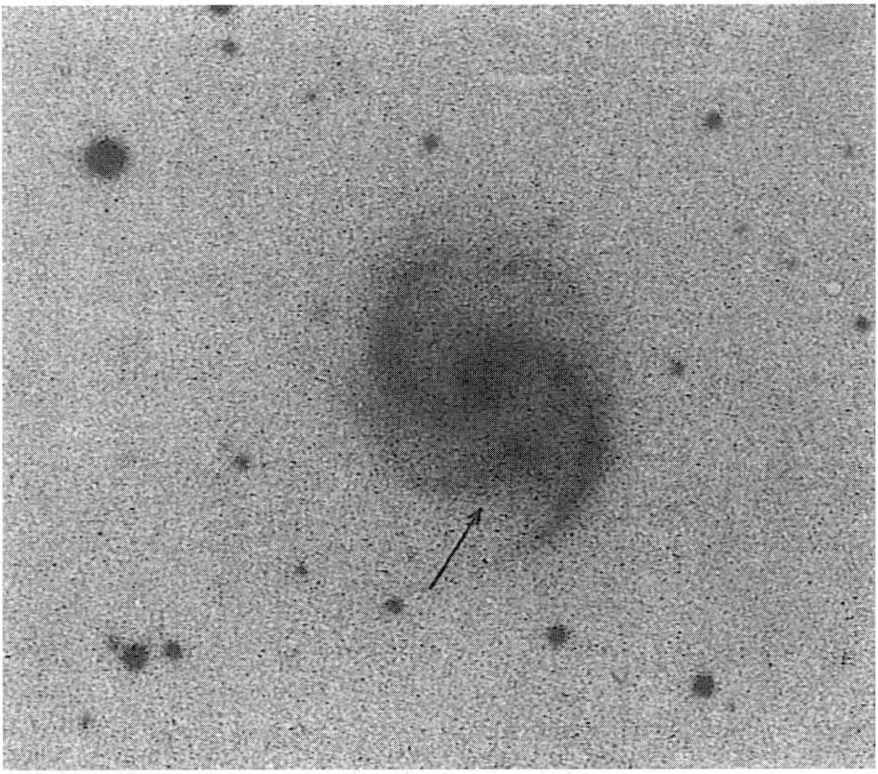

AM 2006-295, a two-arm open spiral system with a curved and elongated galaxy situated within one of its spiral arms. There is a redshift differential of some 22,350 km/s between them. (Photo courtesy of Halton C. Arp)

Discordant Redshift Association

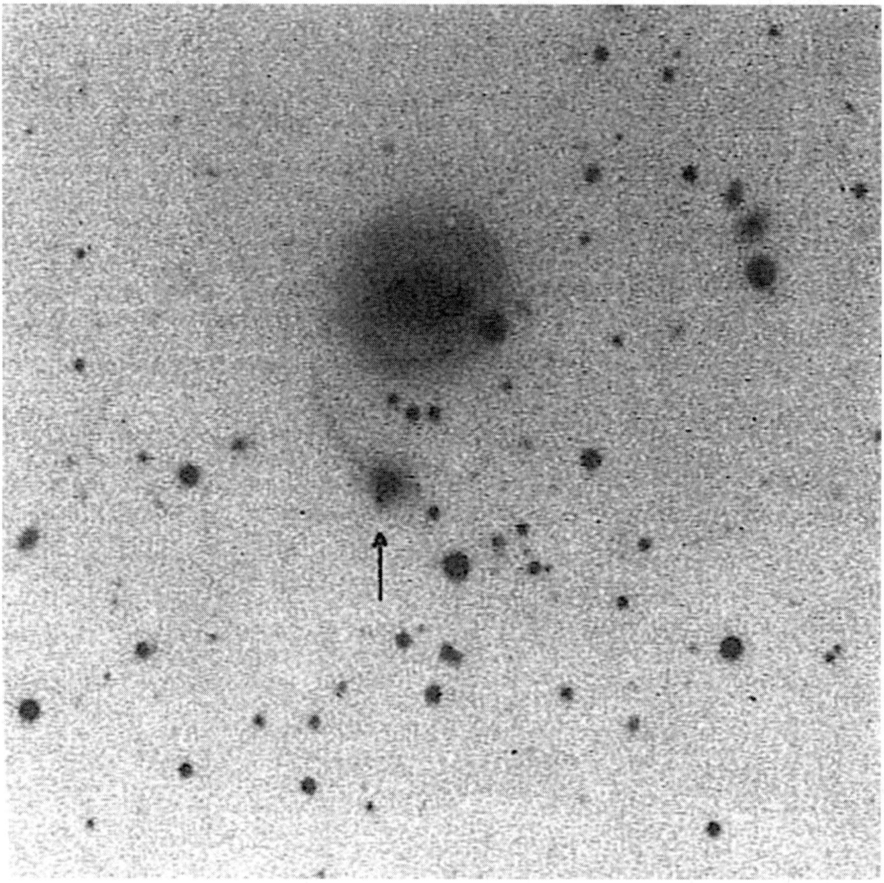

AM 2054-221 appears to be a closed-ring spiral galaxy that has pulled a band of stars and gas from a smaller passing galaxy. There is a redshift differential of some 36,460 km/s between the two systems. (Photo courtesy of Halton C. Arp)

Part Two:

Religious Revelations

7

Biblical Credibility

Inspired Prophets/Priests?— Old Testament Authorship— New Testament Authorship— A Legacy of Unfounded Doctrine— Biblical Symbolism

Tradition frequently leaves deep mental scars. All too often misconceptions implanted in childhood persist throughout the lifetime of an individual, effectively thwarting the revelations of logic and science. Nowhere is this more true than with the exposure of young minds to erroneous biblical concepts. It is, in fact, little short of incredible that widespread ignorance of this body of literature should still persist today in this ostensibly modern era, in spite of numerous and obvious contradictions with both science and reason.

To those whose religious background would deny such valid principles as evolution and reincarnation, it is perhaps expedient to remove any vestige of doubt by going straight to the difficulty. For never, in all the annals of history, have so many souls been so misled by the written word of man.

Inspired Prophets/Priests?

Essential to a proper understanding of biblical scripture is the realization that these documents are not the result of any special revelation, stemming directly from God and bestowed most suspiciously upon but a few chosen individuals. On the contrary, for while there are indeed instances of inspiration, in a great many respects their opinions must be considered quite inferior to the views of modern man with his superior grasp of science. Nor can it be said that works like Genesis contribute anything useful with regard to the more basic mysteries surrounding the creation of life and the universe. Instead, the Old Testament writings depict a lengthy and often very painful struggle of the Jewish race to develop their theology— embracing the concept of a monotheistic God— in the face of hostile neighbors with multiple deities.

Those occasions where God is supposed to have "spoken" to various select recipients must submit to reason. Clearly, incidents ascribed to the actions and commands of God are merely those which appealed to the so-called prophetic writers. For in their trying endeavor to lead the less spiritually evolved masses further along the road to Utopia, they could evidently see no harm in expressing— frequently by means of symbolism or in the name of God— certain ideas and events which they wished to convey. In an era where the masses were prone to accept authority and were unaccustomed to thinking for themselves, this ploy was generally quite successful.

As the unfounded and nebulous dogmas of so many churches reveal today, such an end has failed to justify the means. It is an unhappy fact that once an incorrect and provisionary viewpoint becomes accepted, tradition will only too likely have the adverse effect of deferring later recognition of the real truth. While this state of affairs may have served a useful purpose with respect to ignorant masses of the past, it is certainly time to admit that the prophets signed God's name to their writings because it then seemed expedient for them to do so. Most assuredly, we of a more enlightened generation must awaken and treasure our conscience above the frequently restricting influence of tradition. With the passage of time, is it not to be expected that many of our modern thinkers must have a concept of reality far superior to the views of the ancient scriptural writers?

In order to assess the respective roles of various Hebrew priests in composing the books of the Old Testament, it is instructive to consider a somewhat nefarious side to their lives. Although they professed to have a pipeline to God, in reality they were in a poor position to expound an enlightening philosophy. With very little incentive to engage in open discussion over issues already decided by tradition, a quest for prestige and wealth soon came to dominate their lives. Perpetrating the colossal hoax of animal sacrifice— through Temple-approved priests only— ostensibly in belief that this act would elicit favor from God, for ages a high priest and his minions were able to exact considerable tribute from a gullible populace. Throughout the course of this archaic literature there was, in fact, a constant power struggle— among rival priestly factions— to secure an exclusive franchise in this lucrative but foolish practice. Along with other clues, this priestly struggle is able to supply a measure of insight into the authorship of certain Old Testament documents.

Old Testament Authorship

Essentially, the Old Testament is a mixture of actual and highly exaggerated history, folklore, superstition and outright myth. Also, as we shall see, it contains an ingenious form of symbolism. Handed down by word of mouth and scattered writings for centuries, it was not until the great prophetic era extending from about 900 BC to 300 BC that much of this material was finally gathered together, by an assortment of editors, to form a connected series of documents or books.

The first five of these books, comprising the nucleus of the Old Testament and known as the Pentateuch— Genesis, Exodus, Leviticus, Numbers and Deuteronomy— consist of four primary documents: **J, E, D** and **P**. A brief synopsis of these historical works follows:

Document **J:**

This Yahweh/Jehovah document is thought by many scholars to be slightly older than **E** (about mid-9th century BC). With an unknown authorship which claimed descent from Aaron, their priesthood was centered in the city of Jerusalem, in the southern kingdom of Judah.

Document **E:**

Worshiping a deity by the name of Elohim, their priests were reputed to be descendants of Moses. This unidentified group of northern kingdom priest/prophets likely resided in the city of Shiloh at the time **E** was composed.

Document **D:**

Comprising the manuscript of Deuteronomy, this document appeared rather suddenly in the year 522 BC, during the reign of King Josiah. Believed to have been written by Jeremiah, and Baruch his scribe, it has much in common with **E**— except that it supported centralized sacrifice in the Jerusalem Temple. (These same writers are also deemed to have been the chief author/editors of six other books: Joshua, Judges, I and II Samuel and I and II Kings. The six books are often referred to as the Early Prophets.)

Document **P:**

Known to scholars as the Priestly document, it was evidently written as an alternative to **J** and **E**, with roots extending back prior to the formulation of **D**. The document itself was likely composed by a group of Aaronid priests, following the Assyrian invasion of the northern kingdom in 722 BC, very possibly during the reign of King Hezekiah (727 to 699 BC).

P is the largest of the four sources of the Pentateuch, being equal in size to that of the other three combined. It expounds a more cosmic and less personal account of the sevenfold creation narrative and the great flood story. Included are sagas of Abraham, Jacob,

Moses and accounts of the exodus. It also contains a sizable corpus of priestly law and covers roughly thirty chapters of Exodus, Numbers and the entire text of Leviticus. Both **J** and **E** tell of very similar stories, but with revealing contradictions.

To understand how these four works became interwoven into the early "Torah," it is well to picture certain historical events which played a prime role in this merger. After the death of King Solomon, the Jewish nation was split into two kingdoms in the year 933 BC: King Rehoboam in the southern kingdom of Judah; and King Jeroboam in the northern kingdom of Israel. What emerged was two priesthoods with very different ideas as to who was the rightful high priest heir. When the northern kingdom fell to the Assyrians in 722 BC, many Israelites fled to Judah— bringing with them their treasured **E** manuscript. In due course, **J** and **E** were combined for the sake of unity. But the most difficult task of editing involved the combining of many duplicate but far from identical stories of **P** with **JE**. This merger is believed, by many scholars, to have been the work of an Aaronid priest by the name of Ezra, subsequent to his return from a Babylonian exile about the year 458 BC. Written originally as an alternative to **JE**, in a stroke of irony **P** ends up by being combined with the very text it was designed to replace. The mystery as to why so many Old Testament stories seemed to be duplicates, and with so many subtle contradictions, was at last clear to biblical scholars once they had separated text into its initial root sources.

A key ingredient in this biblical detective case was the way in which the authors of **J, E, D** and **P** favored certain kings and priestly laws— especially laws relating to the idea of centralized Aaronid sacrifice. For instance, King Solomon and King Hezekiah are depicted as tyrants by **E**; while **P** praises both of these early kings, chiefly because of their support of the Aaronid priesthood. Also to be mentioned, by way of distinguishing authorship, is the fact that while **JE** saw no harm in introducing stories of talking animals and glowing angels in order to impart a message, **P** was opposed to such concepts and made no mention of these obvious fabrications. It is a credit to the diligent work of a number of contemporary biblical scholars that the false notion of a Mosaic authorship of the Pentateuch has finally been laid to rest— after persisting for more than two-and-a-half millennia! [1]

On the whole, it may be concluded that the various Old Testament documents were written by a series of priests with two primary ideas in mind. First, they wished to expound their own personal views and aspirations. Second, they desired to convince an uneducated and naive populace of their superior priesthood status— a trait not exactly absent in this present age.

New Testament Authorship

The very notion that a mere human being could really be an actual manifestation of God can only be described as childish, and has all the aspects of a story based upon sheer fantasy and "fairy tale" magic. It simply does not afford a realistic picture of the dynamic cosmos now revealed by modern science. It fails utterly to consider the vast multitudes of other star systems, with their myriad habitable planets, which must exist in a vast celestial universe. Do these other worlds also have a "Christ" such as ours? Do our highly esteemed theologians even know what they are talking about in their support of such a wild idea— a concept which originated centuries ago and in an era which could conceive of nothing beyond a solitary flat Earth and a nebulous "heaven"?

In sticking to plain historical fact, and avoiding the temptation to interpret literally certain biblical narratives that were clearly designed to convey a moral or principle, we may attempt to place the various New Testament writings into proper perspective. Considering that the traditional canon consists of some 27 books, it will be necessary to restrict comment to the main documents comprising this body of literature. (At one time there were as many as 50 or more "gospels" in circulation. It was not until about the middle of the 4th century AD that the present canon was decided upon.)

Jesus himself is now believed by historians to have been born between 6 and 4 BC. According to the biblical records, which is all we have to go by, he was crucified during the governorship of Pontius Pilate in Judea (26 to 36 AD). If we accept as fact the mention in the Gospel of Luke (Luke 3:1), where it is stated that Jesus began his preaching in the 15th year of the reign of Tiberius Caesar, a ministry of one year would put his death at about 29 AD. A possible ministry as long as four years would serve to extend his life until around 33 AD. In any event, there is every reason to suppose that his teaching came to an abrupt and unfortunate end not later than this time, and at the hands of Roman soldiers acting under the orders of Pilate.

The oldest written biblical account dealing with Jesus appears to be the various epistles of Paul (dated from 50 to 60 AD)-- being the works of a man who had never seen Jesus and was not converted from the role of a persecutor until many years after Jesus' death. (Although some scholars hold that I Peter could date back as far as the early 60s AD, it involves the assumption that it was indeed written by the apostle Peter. But since this short epistle imparts little more than encouragement to keep the faith, one must wonder how this relatively trivial document survived while other— presumably more important— works by the same author do not exist. Thus it is likely to be the product of an unknown who was desirous of expressing his views under the name of one more prestigious than himself, and with a date of the order of late 1st or early 2nd century AD.)

The four popular gospels (Matthews, Mark, Luke and John), which are purported to tell the story of Jesus, were all written decades after the crucifixion. The Gospel of Mark (early 70s AD) is widely regarded as being the oldest; while Matthew is probably next, written in the 80s AD. Luke is now thought by some to be dated early 2nd century. The Gospel of John is believed by many scholars to have first appeared on the scene sometime in the 90s AD, long after the apostle by this name must have passed away. Mark was not one of the original disciples, only entering into the picture somewhat later, allegedly as an interpreter of Peter. The work attributed to Matthew is evidently the product of an unknown author who secured much of his information from the Gospel of Mark. (An eyewitness would not have had to depend upon a previous work.) Luke is reputed to have been a companion of Paul and, since neither knew Jesus, would likewise have had to obtain information from an outside source— presumably, from the earlier record of Mark. (There is good reason to suspect that this gospel was written by an unknown who chose to use the name of Luke.) John's version differs in certain respects from the others, being more concerned with the spiritual aspect than any historical Jesus. (It has been determined that another gospel— designated as **Q** by biblical scholars and predating all four canonized gospels— once existed, and that many "sayings of Jesus" were gleaned from this lost manuscript.)

The Acts of the Apostles, ostensibly authored by the same Luke of the third gospel, has been dated at possibly early 2nd century AD. Yet another major work of the New Testament, the Book of Revelation, also belongs to this same period of time. Using the name of John, it is the work of an unknown scribe. An assortment of letters, dated for the most part well into the 2nd century AD, clearly reflect dogmas of the Jesus-Christ myth, which by now had become firmly established.

In any case, the bottom line is that all of the cherished biblical records of Jesus turn out to have been written by persons who had no firsthand knowledge of their subject! (Indeed, II Peter is conceded, by most authorities, to have an antiquity no earlier than about 150 AD— well over a century after the crucifixion!) Thus it will be seen that ample time and opportunity existed for a host of new and totally foreign ideas to have arisen before the teachings of Jesus were committed to scripture— including interpolations of many later authors— and subsequently incorporated into church canon. During this interval there is good reason to suspect that such drastic changes in doctrine occurred as to literally rewrite the philosophy taught by its alleged founder— *changes to the extent that Jesus himself would have been astounded and even shocked by the end result!*

A Legacy of Unfounded Doctrine

A fundamental problem which now arises concerns the issue of Christian origins, in which Jesus the man became transformed into a combination Messiah and Redeemer— to say nothing of equating a human being with God himself! While it must be obvious that neither Jesus, nor his close friends and associates, really believed that he was a divine entity, with all the supernatural powers now attributed to him, how are we to explain the seemingly contrary views expressed in the Bible and long since held by Christian churches to be actual fact?

To understand why the New Testament appears to support a Messianic philosophy we must realize that, at the time of Jesus, the Jewish people lived in constant expectation of a Messiah who would come and establish a new world order. Subsequently, ii will be seen that, following the death of Jesus the teacher, in such an atmosphere it would be only a matter of time until advantages were to be had in the preaching of a Messianic gospel. Since past experience with so many Old Testament writings had shown that acceptance of a particular doctrine by the masses was largely dependent upon the prestige of the author, it was most desirable to attribute the highest possible prestige to the master whose works they wished to propagate.

Thus, while the idea of portraying Jesus to be a Messiah may well have been unforeseen before his death, his martyrdom now offered a great opportunity to claim a divine origin and thereby obtain much valuable prestige and authority. As to the question of whether or not Jesus deliberately laid down his life for the sake of Principle, we can only surmise as there would appear to be two possibilities. On the one hand, since his teachings were so opposed to priestly practices at the time, we must consider that he was not unaware of the dangers involved. If this is so, then we may wish to look upon the crucifixion of Jesus as truly a deliberate act of unselfish love, motivated by conscience, for the benefit and example of all mankind. On the other hand, (as will be discussed later) there might well be another side to this picture.

In any event, in view of the antiquity of authorship and lack of genuine firsthand information, along with the inevitable "editing" by a series of opinionated writers, it will be seen that the task of incorporating later theological ideas was relatively simple. Indeed, in this respect they were aided by a favorite custom of biblical writers in general who were inclined to word certain stories so that they were capable of a "dual" interpretation. That is to say, if it was to their advantage they liked to combine symbolism with text that was conducive to a literal interpretation, if readers were willing to give it such credibility. It is, therefore, not surprising that many false ideas were accepted in the course of time and that the status of Jesus is marked by the observed element of confusion.

Invariably, one of the most important of these false teachings to plague the early church is that of the doctrine of redemption. This belief is somewhat peculiar and apparently found little sympathy among the first Christians of Palestine, as is evidenced by the fact that it did not develop until the new religion had spread out into the pagan world. The Messianic view held by the Jews at the time of Jesus did not center at all upon the idea of Messiah who would be a Redeemer making atonement for any "Original Sin" dogma. There seems little doubt but what the weird Redeemer concept owes much to pagan beliefs of the time.

Mithraism, in particular, definitely influenced early Christian thought to a considerable extent. This ancient religion, which originated in Persia as only one of a number that embraced the principle of a Redeemer or Mediator between God and mankind, was firmly established in Rome at least a century before the crucifixion. It flourished until the 4th century AD, when it finally succumbed to the new and highly competitive religion called Christianity. However, this surrender was evidently far from unconditional, since several features of the older religion were clearly absorbed into the structure of its successor— *including the very practice of observing Christmas upon the actual birthday of Mithras himself!* Other supposed Christian practices derived from Mithraism include switching the Jewish Sabbath to Sunday, the first day of the week, which had long been dedicated to the Sun and was the chosen holy day of Mithraics. This religion also contributed such ideas as baptism and confirmation, and fostered notions of a mystical salvation from a eucharistic Last Supper.

Until this fateful merger with Mithraism, some centuries after the death of Jesus the teacher, there is reason to believe that many of the early church members accepted the principle of reincarnation as a viable philosophy. Indeed, since it was not an unpopular belief during the lifetime of Jesus, it is instructive to note that *at no time did he see fit to repudiate it!* On the contrary, such references as "In my Father's house are many mansions" (John 14:2) and "You are of more value than many sparrows" (Matthew 10:31), tend to suggest the principle of reincarnation and (perhaps?) even a Cosmic Pyramid. In any event, history reveals that among the more prominent of the early Christians who embraced the concept were Origen, St. Augustine and St. Francis of Assisi.

If there was one major contributor to the idea that Jesus was a divine incarnation of God it was Paul, an educated Jew from Tarsus. Subsequent to his conversion, this dynamic apostle was sent out of Jerusalem to neighboring countries on a mission to preach the wisdom of Jesus. It was evidently not very long before he saw merit in elaborating upon the works of his idol, in order to more successfully compete with other religions and their semi-god deities. In just a short while, Jesus the teacher was transformed into Jesus-Christ the Son of God. Once this great leap had taken place before one audience it soon became impossible to retract such exaggeration, and a new and highly revolutionary religion was born— for better or for worse. (As we shall see, this unfounded claim might well be at the root of Paul's later conflict with James, a prominent leader of the early church in Jerusalem.)

Thus it came to pass that many confused souls, desperate for guidance and leadership, fell prey to attractive promises inherent in the new faith called Christianity. In a stroke of irony, with the passage of time even the mighty Roman Empire succumbed to the very religion which it had previously sought to suppress. Spreading from such a strong focal point, it was not long before the concept of a centralized leader was adopted as supreme authority of the Roman Catholic Church: the pope.

Once endowed with the power of an absolute monarch, it was

merely a question of time until a host of dogmatic ideas were conceived and thrust upon a populace that was coerced into submitting to priestly authority. Respect for an individual's conscience and freedom of thought were eventually ignored— culminating in acts of outrageous persecution during the infamous inquisitions of the Middle Ages. In compliance with the adage that "absolute power corrupts absolutely," the pinnacle of church folly was attained in the year 1870 when the absurd doctrine of "papal infallibility" was promulgated. In essence, the pope had now become God in matters of faith and morals!

Rebellion against the Roman Catholic Church took a rather bizarre twist as early as the 7th century AD. Originally holding sincere feelings toward Christianity, an Arab by the name of Mohammed became disenchanted by the disunity and bickering among rival church factions and decided to found his own version of this religion. By the time of his death (632 AD) the new religion of Islam had gained a firm foothold in the Arab world and was poised to spread to adjoining nations. In yet another example of irony, the Catholic Church had inadvertently spawned a new and highly competitive religion— one which would likewise seek to suppress thought and impose its convoluted beliefs and practices upon all those falling under its sphere of influence.

Invariably, there is a vital lesson to be learned from the history of organized religion, with its strong emphasis upon dogma that would forbid criticism and discourage further reflection on the grounds that ecclesiastical authority knows best. But a philosophy based primarily on emotion and blind faith can only serve to deceive the believer. To obtain a more realistic perspective we must free ourselves from all vestiges of priestly tradition and ponder the universe through the eyes of an open-minded scientist.

Biblical Symbolism

That the vast majority of both Old and New Testament stories are not to be regarded as factual, from the point of view of historical accuracy, becomes quite evident once we examine the rather unique system of symbolism which pervades these ancient writings. For undoubtedly, the most obvious and most striking characteristic of the Scriptures is to be found in pronounced repetition of certain "sacred" numbers, cryptographic phrases and words. In fact, when one contemplates the exceedingly lavish use of numerals, it will be noted that practically every such occurrence may be traced to one of a half dozen numbers, or at least to their multiples or to special combinations! Why should this be so? Therein lies a most important and largely obscured facet of biblical literature; for the writers of these documents were concerned with much more than actual history and mere physical events. To them, a chief aim was to convey an intended *moral or principle*, quite regardless of whether there was any correlation with reality or whether mythology and fiction had to be resorted to in order to stress the significance of a point.

This they chose to do by devising a system whereby certain names, persons, places, objects, words, phrases, happenings—and especially *numerals*— were introduced for the specific purpose of *representing or symbolizing certain spiritual ideas or principles*. For instance, in connection with the extensively used system of numerical symbolism, the basic numbers 1, 3, 4, 7, 10, 12 were adopted and given a special meaning. A brief insight into these subsequently idolized "sacred" numbers follows:

One:

The number "one" was held tin high esteem as a symbol of the recognition of a solitary *Supreme Being*— that there is in reality but one God ruling the universe. (The realization that there is only one true God that mankind should worship was quite a bold step in those days.) An example would be, "could ye not watch with me one hour?" (Matthew 26:40).

Three:

The figure "three" came to be a rather important symbol which was used to represent the *intimate nature of spiritual matters, of success in realizing or attaining some measure of Truth.* The adoption of this particular numeral may well have been suggested by the overall sum of the conspicuous heavenly bodies— namely, the Sun, the Moon and the stars. A few of the many instances of the usage of the numeral "three" would be Jonah in the whale's belly three days and three nights, the rising of the "Christ" on the third day and the episode of Saul (later renamed Paul) blinded for three days while on the road to Damascus.

Four:

The symbol "four" was probably derived from the four cardinal points of the compass and perhaps also from the four fundamental processes of mathematics: addition, subtraction, multiplication and division. Pertaining to spiritual values, this number is used to signify *a sense of direction and spiritual contemplation— the act of seeking the Truth* (e.g. the "four winds" mentioned so many times in the Scriptures, the four who set out for the land of Canaan in the Abraham narrative, the four thousand fed by Jesus, the raising of Lazarus who was dead for four days, etc.)

Seven:

The number "seven" is accorded a high and prominent position in the Scriptures. Apparently derived from the sum total of those heavenly bodies known to exhibit motion in ancient times (Sun, Moon, Mercury, Venus, Mars, Jupiter, Saturn), this figure came to represent the completeness of the heavenly system. Hence, this number was used to symbolize a *sense of divine fulfillment and of spiritual completeness or perfection.* (In addition, as will soon be revealed, the number "seven" was also used to convey the spiritual meanings of the seven days of creation story.) A few examples of the use of this number would be the seven days of creation, Moses' seven ascents of Mount Sinai, the numerous sevens appearing in Joshua and which concern the falling of the walls of Jericho, the fact that (in Luke's genealogy) Jesus is depicted as the 77th personage, the seven messages, the seven seals, the seven trumpets, the seven vials mentioned in the Book of Revelation, etc.

Ten:

The adoption of the figure "ten" as a symbol is believed to stem from the total number of our fingers, upon which we depend to perform so many human tasks. Thus the important symbol "ten" was selected to depict the actual *application of spiritual thought, of striving to put one's abilities and potential to work.* Examples would be the Ten Commandments, the parable of the ten virgins, Jacob's solemn statement (Genesis 28:22) that "I will surely give the tenth unto thee," etc.

Twelve:

The symbol "twelve" is evidently derived from the twelve signs of the zodiac, which at one time were held to influence and determine the actions and fate of man. Consequently, the numeral "twelve" was adopted to symbolize a *demonstration of true government, of the authority and universal extent of divine principle* (e.g. the twelve sons of Jacob, the twelve stones carried across the River Jordan, the twelve apostles, etc.).

While the above "basic" numerals are by far the most common and are used generally in the senses described, there are a few exceptions which deserve mention. For instance, the numeral "six" is often used as a symbol of mortal and imperfect thought. Another occasion would be the use of a particular numeral pertaining to the message imparted by one of the seven days of creation (e.g. the number "two" could refer to the message conveyed by the second day of creation, "five" to the fifth day, etc.). In addition, there are also times when the symbolic role of these "sacred" numbers is reversed and used in the *opposite sense!* Typical examples of this usage would be the three temptations of Jesus by the devil, Peter's three denials, Judas receiving three times ten or thirty pieces of silver for betraying Jesus, etc.

Once it is finally recognized that the Scriptures were intended to impart a spiritual lesson or point of view, rather than record mere physical and historical occurrences, they immediately take on a new and higher meaning— the picture at last begins to make sense. That this most conspicuous aspect was ever doubted is quite surprising in view of the remarkably consistent and obvious usage of symbolism of all forms. Much more astonishing is the fact that so many religious leaders still fail to realize, even today, that such stories as the seven "days" of creation, the Adam and Eve narrative and the various "miracles" were devised for the purpose of illustrating a particular lesson. (Along with motivation of a spiritual nature, which is generally of fundamental importance, it is to be noted that some passages were written with a secondary purpose in mind. That is to say, they were designed in such a manner as to actually lend themselves to a literal interpretation— if simple minds would be willing to give them this credibility.)

In a way, we can perhaps understand and partly forgive the actions of the ancient writers, since they must have had a strong incentive to provide the inquisitive masses with some tangible idea of how the world came into being. But this excuse is hardly valid in our present age of enlightenment, and those who would continue to preach a literal interpretation in the face of reason are in reality false prophets doing much to hinder the advance of truth.

References:

[1] One informative reference source which may be cited is *Who Wrote the Bible?* by Elliot Friedman, ISBN 0-671- 63161-6.

8

The Old Testament

The Seven "Days" of Creation— The Adam and Eve Story— Noah and his Ark— Some Early Biblical Narratives— The Saga of Moses— So-called "Miracles"—Summary

Armed with insight into the aforementioned system of symbolism, we are now in a position to make sense out of a host of otherwise obscure and enigmatic biblical stories. In the process, it will be seen that many characters and events have not the slightest basis in fact, and were invented primarily for the purpose of conveying a specific moral or concept which various priest/writers wished to express.

The Seven "Days" of Creation

Seizing upon a rather similar Babylonian creation myth, certain writers proceeded to develop it as an introduction to the Adam and Eve story, which had been composed quite some time before. When we recall the symbolic meanings attached to the various numerals, it soon becomes apparent why their creation narrative happens to be extended over a period of seven days, since the figure "seven" is a symbol of fulfillment and completeness.

Although designed to give simple minds an explanation of how our physical universe came into being, in reality the seven "days" of creation may be resolved into a unique symbolic presentation of an instructive moral or principle. Depicting a sevenfold process of *evolving spiritual thought*, these stages may be stated as follows:

Day One:

The first recognition of ***conscience***, of an Intelligence or Mind controlling the universe (i.e., the *light* of the first day of creation, since the Sun, Moon and stars have yet to be introduced), is expanded systematically through:

Day Two:

A fuller understanding of the significance and capabilities of conscience, and the sensing that *it is able to **separate** good from evil and lift us above darkness* (i.e., the firmament or *separation* of the second day);

Day Three:

The great spiritual uplifting to be had from following conscience must bring to one a ***tangible sense of reality*** *which can ever be relied upon with assurance* (i.e., the *dry land* of the third day);

Day Four:

The realization that the revelations of conscience are the outcome of a divine ***System or Principle*** influencing mortal thought, and capable of leading one in the direction of Truth (i.e., the divine system of the celestial universe epitomized by the *Sun, Moon and stars* of the fourth day);

Day Five:

A great sense of ***exalted thought and abundance of life*** due to the progressive revelations made by conscience (i.e., the fifth day where the *fowl of the air* are used as a symbol of exalted thought, and the *fish of the sea* symbolize abundance);

Day Six:

The recognition that man is to have dominion and that he is the ***ultimate*** earthly creation and that his soul is ***immortal*** in nature, since he is capable of intimate communion with God (i.e., the sixth day on which man is created in the *"image"* of God);

Day Seven:

Finally, the realization of the wondrous potential possessed by man must bring to him a great sense of ***security, peace and rest***, since the attributes of *immortality* and of *communion with God* are all that could be desired by one (i.e., the *completeness, fulfillment and state of rest* epitomized by the seventh day).

An understanding of the underlying message imparted by each of these seven "days," or stages of spiritual thought, is essential to a proper interpretation of the Scriptures. Indeed, the essence of these stages of spiritual development is embedded in numerous biblical passages.

The Adam and Eve Story

Immediately following the seven days of creation story comes a narrative purporting to give an explanation for our imperfect environment. That the Adam and Eve story cannot possibly receive a literal interpretation is now so obvious as to be beyond question, both from the aspect of historical accuracy and from an ethical point of view. To accept a literal translation of this narrative is to deny free will altogether; *it is to cast blasphemy upon the just nature of the spiritual cosmos!*

It is as strange as it is disturbing that religious groups have been able to perpetuate a sadistic hoax for so long. It is strange to think that so many supposedly intelligent theologians could preach an obvious untruth, and highly disturbing to find that there are so many who would defy common sense and listen to them. To think that our imperfect environment is the result of the sin of one man is more than fantastic; *it is positively unthinkable!* On the basis of a literal interpretation we are not the children of a Just and Loving Creator, but the unfortunate victims of a God of spite! To even insinuate that we should be made to suffer for the sins of a man whom we have never encountered is just about the most shocking belief imaginable. For if we are responsible for the sins of those who lived before us, then there can certainly be no such thing as free will— life itself becomes a complete farce! What, then, is the true value of this narrative?

As is typical of so many of the early biblical stories, none of the characters portrayed in this particular allegory ever existed in real life, but are fictitious symbols introduced for the purpose of conveying a moral or lesson. In direct contrast to a literal interpretation, the real meaning of this story was to signify that man's sense of conscience was of a distinctly higher nature than that of all other forms of earthly life. In reality, the Adam and Eve story was written for the express purpose of illustrating the principle of *conscience and free will*— certainly not to deny them, as a literal interpretation must imply!

Upon looking over the text of this plot, the evidence in support of a symbolic explanation is so obvious that it is amazing how anyone could have thought otherwise. First of all, is it not of the greatest significance to note that the tree from which Adam and his wife ate, and through which action our imperfect environment is supposed to have originated, happens to be so aptly termed "the tree of the knowledge of good and evil" (Genesis 2:9)? Indeed, is not the symbolism used to illustrate the recognition of conscience perfectly clear when we read, "For God knows that when you eat of it your eyes will be opened, and you will be like God, knowing good and evil" (Genesis 3:5)? And furthermore, what were the immediate consequences of eating from this magical tree? "Then the eyes of both were opened, and they knew that they were naked" (Genesis 3:7).

Thus the reason why Adam— *a fictitious symbol representing mortal man*— is caused to suffer as a result of his actions is seen to be quite natural enough, since a greater knowledge of God's Will must necessarily entail exposure to abuse and persecution from his fellow beings who are less evolved and who do worship material rather than spiritual values. This point, incidentally, is promptly portrayed in the fictitious saga of Cain and Abel— where the lowly physical sense slays one of higher principles— as well as in several other narratives throughout the Scriptures,

Noah and his Ark

Like the symbolism characterizing the seven days of creation and the Adam and Eve story, the extraordinary occurrences that are reported in so many biblical accounts were similarly evoked for the specific purpose of imparting a spiritual lesson.

For instance, take the episode of Noah and his ark. Who in all honesty can accept this story as an actual historical event? On the basis of a literal interpretation it is clearly absurd; the story breaks down completely from every conceivable angle. It is not only plainly ridiculous from an historical perspective, but a deliberate deluge of such destructive proportions is wholly inconsistent and unacceptable from an ethical point of view. Invariably, all attempts to interpret this allegory in a literal fashion encounter so many insurmountable difficulties that one would expect this passage to serve as an incentive to bring theologians back to reality— back to the realization that the authors of the Scriptures were concerned primarily with *illustrating principles by means of symbolism!*

Actually, no such personality as Noah ever existed. Like the mystical Adam and his descendants, Noah is simply another fictitious character (or symbol) introduced for the purpose of expressing a spiritual lesson or moral. As a matter of fact, the symbolic aspect of this narrative is so conspicuous that it would have to be regarded as an allegory— even if the story itself was within the bounds of reality. Is it not somewhat of a coincidence to read that Noah's father, Lamech, happened to live to the rather strange and unusual age of *"seven hundred and seventy-seven years"* (Genesis 5:31)? Is it not also significant to further read that Noah had *three* sons (Ham, Shem and Japheth), that the length of the ark was *three hundred* cubits (the multiplied sense of the "ten" and "three"), the breadth of it *fifty* cubits (the meaning of the fifth day of creation combined with that of the numeral "ten") and the height of it *thirty* (or "three" times "ten") cubits (Genesis 6:15); that the ark should contain *three* stories; that Noah was *six hundred* years old when he entered the ark (the sense of the sixth day of creation combined with that of "ten"); that it rained *forty* days and *forty* nights (the symbolical meanings of the numerals "four" and "ten" interwoven together); that the ark rested in the *seventh* month, on the *seventeenth* day ("seven" plus "ten") of the month, upon the mountains of Ararat and that in the *tenth* month, on the *first* day of the month, were the tops of the mountains seen (Genesis 8:4-5); and that Noah sent out a dove (a symbol of purity and peace) *three* times, at intervals of *seven* days? Indeed, when we also consider that throughout biblical literature the *ark* is held to represent a high and inspired mode of thought, that the *pitch* stressed in its construction is a symbol of atonement or 'oneness" with God, and that even the very word "Noah" itself means "rest" or "comfort," does it not become apparent that the whole affair is nothing more than a symbolic presentation of some cleverly conceived moral or principle?

The overall message to be drawn from this obvious allegory may be defined as follows: Whenever we find ourselves engulfed by the floods of materialism and despair, we may find safety and solace by casting our eyes in the direction of God; in response to our efforts He will lift us above the ravaging effects of mortal actions and experiences.

Some Early Biblical Narratives

Bearing in mind that the ancient prophetic writers were more interested in matters of spiritual importance than they were in mere historical facts, let us consider briefly the significance of certain early biblical characters and narratives.

While the purely mythical characters of the Old Testament do give way eventually to actual personalities (even if somewhat exaggerated and idolized), there is a lack of tangible proof with regard to the reality of many earlier personages. However, once it is realized that certain individuals were introduced into the Scriptures for the purpose of illustrating specific morals, it will be seen that it makes little difference whether they are fictitious or not, since the underlying principle is of far greater importance than a mere physical and transitory event.

Abraham:

The character of *Abraham* is introduced into the Scriptures for the purpose of portraying a more advanced understanding of man's true relationship with the Father. As opposed to a deep-thinking Abraham, we have the *Lot* state of thought which might be said to display a worldly and somewhat indifferent attitude toward God— namely, one who hesitates to expend the necessary effort to evolve spiritually. One other participant in this narrative is Abraham's wife, *Sarah,* who is injected as a symbol of the necessity to keep one's thoughts "wedded" to matters of a spiritual nature.

The story itself gets underway with mention of *four* people setting out for the land of Canaan. Understanding the significance of the numeral "four" and the subsequent command to "Go from your country and your kindred and your father's house to the land that I will show you" (Genesis 12:1), can we not see that here we are dealing , not with an expedition to some foreign country, but with *an expedition setting out in search of Truth?* Have we not heard, at one time or another, this same call to "Go from your country"— to abandon the enslaving ties of all self-centered and earthly aspirations (including the utterly foolish sense of nationalism) and strike out bravely in search of the freedom and true glory to be found in a superior philosophy?

On the rough and lengthy road to an advanced state of mind it is recorded that Abraham went "down to Egypt to sojourn there" (Genesis 12:10) and that he promptly got into trouble through falsely stating that Sarah was only his sister. Recalling the symbolic meaning attached to the name of "Sarah," along with knowledge that "Egypt" is often used as a symbol of unenlightened thought, one is enabled to arrive at the following interpretation: In the face of trials and tribulations, and when adherence to our "Sarah" (or obligation to seek the Truth) might bring us into disfavor, are we not all tempted to disguise any unorthodox views and make believe that we are not so closely attached to such beliefs as we really are— that it is not our "wife" but only our "sister"? However, Abraham's sense of reality was too high to permit him to stay long in Egypt, and it is reported that he "went up from Egypt" and "called on the name of the Lord" (Genesis 13:1,4)

As is bound to be the case when an individual's perspective has increased to a point somewhat in excess of those around him, and if remaining together would only hinder advancement, then a parting of the ways must take place. Knowing full well that Lot would choose the way of spiritual stagnation, Abraham felt safe in offering Lot the choice of turning either to the right or to the left. Consequently, while Lot chose the easy way— the "Jordan valley" that was "well watered"— and eventually had his very beliefs and foundation destroyed under him (the symbolic destruction of Sodom and Gomorrah), Abraham continued along the rugged upward path to the "land of Truth." Our obligation to seek the Truth (i.e. the command to look in all *four* directions), and the increased clarity to be had once we divorce ourselves from adverse influences (or the "Lot" state of mind) is well illustrated by the following: "The Lord said to Abram, after Lot had separated from him, 'Lift up your eyes, and look from the place where you are, northward and southward and eastward and westward; for all the land which you see I will give to you and your descendants for ever'" (Genesis 13:14-15).

And yet, spiritually minded as one might be, there are very often temporary lapses. This necessity for man always to struggle against his own slothful nature is stressed in the episode of Abraham forsaking Sarah and taking a mere maidservant (Hagar) to wife, the result of which union was *Ishmael*, who symbolizes the everyday physical aspect of mankind. Finally, Abraham's sense of values rose sufficiently to make him understand that the "Ishmael" state of mind must be cast loose, regardless of how much we might wish to retain it. That stories such as this are allegories, and were never meant to receive a literal interpretation, is further borne out by the reference of Paul to this particular saga in Galatians 4:22-26, in which he commented on this story: "Now this is an allegory!"

Isaac:

Although obliged to omit many interesting incidents relating to Abraham, there is perhaps one more regarding the introduction of the symbol *"Isaac"* which should be mentioned. It is stated that *three* men of the Lord appeared at the door of Abraham's tent and told him that a true son or heir would be given unto him through Sarah. Recalling the meaning of the numeral "three," along with the knowledge that Abraham represented an advanced sense of man's true relationship with the Father, this allegory might be translated as follows: Abraham came to realize that his more advanced state of thought could never perish, but must live on and multiply in the minds of others.

Consequently, with the death of Abraham, Isaac— his promised son— now inherited the role of an advanced spiritual thinker pressing ever onward in search of Truth. Thus we are enabled to grasp the real significance behind Abraham's request that Isaac never return to the country which he himself had left, since one in genuine search of the Truth can never turn back but must continue to forge ahead. (Witness the symbolic episode of Lot's wife who was turned to "salt" when she looked back.) When we recall how Sarah was used as a symbol to Abraham, it is not surprising that *Rebekah* became a similar symbol to Isaac. Further, it is rather interesting to note that "Isaac was *forty* years old when he took to wife Rebekah" (Genesis 25:20).

Jacob:

Just as in the Abraham saga, where an advanced perspective is contrasted with a "Lot" state of mind, so in the Jacob narrative we have a similar parallel with a "*Jacob*" sense (a sort of Abraham and Isaac) and an *"Esau"* sense (so typical of the average individual). The first significant incident regarding these two contrasting figures lies in Esau's reported selling of his birthright to Jacob for mere physical food. What does this selling of one's birthright signify? It signifies that the Esau state of thought, when put to a test, will be only too likely to succumb and relinquish its "birthright"—that of spiritual insight of lasting value—for some temporary physical gratification. How frequently do we all tend to loosen our grip upon our "birthright" for the sake of obtaining advantages which are of but a temporary and material nature!

Shortly after this selling of the birthright and the subsequent blessing of Jacob by Isaac, his father, it is reported that the jealous wrath of Esau forced Jacob to leave home. (Is it not characteristic of mortal man to become jealous of another's success— especially when one starting life in less favorable circumstances is observed to pass us and gain possession of something desirable which could have been ours had we been but willing to expend the necessary effort?) Jacob evidently deemed it advisable to flee the wrath of Esau because he felt that his perspective was not yet sufficient to deal with really pressing tribulations. He needed time to construct a philosophy capable of overcoming the negative influence of an "Esau" state of mind.

Accordingly, after Jacob fled from Esau, it is recorded that he spent the night at a place called Bethel, which means "the house of God." Moreover, it is stated that after "Taking one of the stones of the place, he put it under his head and lay down in that place to sleep" (Genesis 28:11). Since throughout the Scriptures the word "stone" is used as a symbol of the *appearance of Truth*, it follows that Jacob was indeed earnest in his quest. The success of this search is stated in the very next verse: "And he dreamed that there was a ladder set up on the earth, and the top of it reached to heaven; and behold, the angels of God were ascending and descending on it" (Genesis 28:12). The ladder reaching up to heaven symbolizes Jacob's understanding of the ordered process of spiritual advancement (i.e., the ascending scale of the seven days or stages of creation); the angels ascending and descending are symbolic of seeking (the ascending angels represent ascending thought) and finding some measure of Truth (the descending angels symbolize the "coming down" or entering of conscience— the *"voice of God"*— into one's mind). The fact that "Jacob rose early in the morning, and he took the stone which he had put under his head and set it up for a pillar and poured oil on the top of it" (Genesis 28:18) signifies that Jacob decided to consecrate his life to the search for Truth, since "oil" is used frequently as a symbol of *consecration.* Also of interest is the statement that "I will give the *tenth* to thee" (Genesis 28:22), for interpreted it means that Jacob planned to put *his increased knowledge into practice.*

Just as in the examples of Abraham and Isaac, Jacob also found it desirable to keep his thoughts wedded to spiritual matters. Hence it is reported that Jacob said unto Laban, "I will serve you *seven* years for your younger daughter Rachel" (Genesis 29:18). But after serving the seven years (or more precisely, attempting to use the "sevenfold path to Truth), Jacob found himself in possession of Leah, the older daughter. While perhaps not too poor of a companion, she was still short of his ideal of Rachel. However, Jacob was not one to give up so soon, and by promising to serve Laban yet another seven years he finally won his "ideal," or higher sense of values.

On the way back at last to face a climax with his brother Esau, Jacob was plagued with a struggle inasmuch as he began to fear that his perspective might still be inadequate to cope with the Esau state of mind. Thus at the brook of Jabbok we find Jacob engaged in a "wrestling bout" with himself, the outcome of which was assurance that his superior vision would see him safely through his darkest hours. This assurance is soon illustrated in the changing of Jacob's name to *Israel*; for where "Jacob" meant only "contender," the name "Israel" is symbolic of a "prince of God." (At a similar crossroad in life Abram had previously been renamed Abraham and Sarai, his wife, was renamed Sarah.) The validity of this opinion was borne out when Jacob and Esau finally met, for while their views remained opposed they abided in peace, physically speaking, and Jacob had naught to fear.

Before concluding the Jacob narrative it is well to mention the significance of Jacob's *twelve* sons. Like all that has come before, this "twelve" is not to be taken literally, but rather is intended to symbolize the universal scope and authority of *divine Principle and true Government*. The "twelve" sons in this instance infer that the advanced Jacob state of mind must multiply and exhibit authority. (The twelve sons of Jacob are supposed by tradition to have formed the twelve tribes of Israel.)

Joseph:

Of the twelve "sons" allegedly fathered by Jacob, it is somewhat instructive to note that only *Joseph* and *Benjamin* are actually born of Rachel, Jacob's "ideal." Jacob's blessing of Joseph meant that this son now inherited the symbolic role of portraying an advanced state of spiritual thought. (The "coat of many colors" which Jacob gave to Joseph represents receiving insight into the meaning of the seven days of creation.) With the birth of the high Joseph state of vision, it is stated that Rachel desired for the Lord to "add to me another son" (Genesis 30:24). What is meant by this strange perception that "another son" must be added? The answer is to be found in the *ability to give up the mortal and worldly sense of things* which the "Benjamin" sense symbolizes; for a higher perspective is of little avail unless one is able to really accept— and live— a superior philosophy of life, with no regard for personal cost. Hence it is reported that Benjamin was brought forth through sorrow and anguish, as it is indeed painful to give up many things of this world that we have come to cherish.

Just as Abraham and Jacob were matched symbolically against opposing and inferior states of mind (the Lot and Esau mentality), so

Joseph had to struggle against his less spiritually evolved brethren. The result was that his brethren, being unable to tolerate a superior and more demanding state of thought, stripped him of his coat of many colors and sold him into Egypt. (In their foolish endeavor to thrust a literal interpretation of the Scriptures upon us, have not our theologians of today been guilty of stripping a "coat of many colors"— the light of Truth— from us? In effect, would they not sell us into the same state of darkness as is symbolized by Egypt?) However, as much as one may attempt to suppress spiritual thought and subject us to trials and tribulations, if our philosophy is built upon a firm foundation, then we cannot help but rise above the situation. This is exactly what Joseph managed to do, for instead of remaining in weak submission he rose to such great prominence that even his brethren had to come to him on bending knee.

The purported incident of Joseph translating the Pharaoh's two dreams (seven thin cows consuming seven fat cows, and seven thin ears of corn devouring seven fat ears) is perhaps deserving of comment. Joseph's interpretation that there would be seven years of plenty to be followed by seven years of famine does not refer to matters of a mere physical nature. On the contrary, the moral which the writers of this narrative were attempting to stress is that, as filling as an understanding may be of the spiritual meaning behind the seven days of creation (the seven fat cows and the seven fat ears), unless we take care to preserve this light, the influence of our materialistic environment (symbolized by the seven thin cows and the seven thin ears— depicting a "counterfeit" sense of the seven days of creation) will destroy our precious light. As Joseph advised the Pharaoh, the best safeguard is to make full use of our opportunities (the seven fruitful years) to build up our storehouses with harvest (spiritual thought) so that in the face of famine or tribulation (the seven lean years) we will have substance to see us through.

That this story is simply another instructive allegory is clearly demonstrated when the starving Egyptians came to Joseph for food *three* times. The first time Joseph took all their money; the second time he took all their cattle, and finally, the third time, he took not only their lands but also their bodies (Genesis 47:14-21). Interpreted literally, one can hardly escape the conclusion that here indeed is a profiteer to end all profiteers. But this is not so, for the "starving" Egyptians were not people suffering from any physical hunger, but rather they symbolize a humanity beset by a *spiritual* hunger. In essence, the food which Joseph gave them was much needed spiritual insight. The fact that Joseph took all their worldly possessions from them (including jurisdiction over their physical bodies) symbolizes that only the spiritual is of real importance, since that is all that was left to them!

During this famine it is recorded that while Jacob sent *ten* sons to buy corn in Egypt, he also declined to send the all-important and symbolic "Benjamin"—namely, the ability to give up the worldly sense of things. But Joseph realized that, valuable as may be an attempt to apply a measure of spiritual insight to human experience, any such action is quite incomplete without the ability to divorce oneself from materialism and worldly tradition. Hence they did not receive sufficient food until Benjamin was brought before Joseph and all twelve sons were finally united (i.e., the "ten" representing the application of spiritual thought to the human problem must eventually rise to the "twelve," symbolizing the recognition and acceptance of divine Principle).

The Saga of Moses

Moses is perhaps among the first Old Testament characters for whom grounds might be said to exist in support of a real personage, although it is apparent that the story of his life has been told so as to express the symbolic polices and aspirations of subsequent Hebrew prophets. Indeed, the biblical record of Moses is so saturated with symbolism that a literal interpretation of many of the reported events is quite out of the question. Recalling that all existing accounts of Moses were not penned until many centuries after his death, and that at the time of the scriptural writers he had come to be idolized by the Hebrews, it will be seen that there prevailed a marvelous opportunity to attribute to Moses almost anything which later editors were desirous of propagating. Just what views were actually those of Moses and which must be ascribed to a succession of religious scribes is a matter of pure conjecture, although judging by the extensive use of the "sacred" numerals, the influence of the Priestly era writers alone is most conspicuous.

Conceding the need to restrict comment to little more than a brief outline of the rich symbolism enveloping the life of Moses, a few of the more noteworthy examples are as follows:

The Early Life of Moses:

The circumstances surrounding the birth of Moses are somewhat similar to those connected with the birth of the babe Jesus, in the sense that both Pharaohs and Herods alike seem to have an uncanny ability to predict the *appearance of light* In both cases the reported slaying of male Hebrew children are not to be taken literally, but instead symbolize that our imperfect world is only too inclined to suppress or slay anyone seeking to reveal the Truth. The tale of the babe Moses being hid *three* months (symbolizing insight into spiritual matters) and then placed into an *ark* (symbolic of a high and inspired mode of thought) daubed with *pitch* (a symbol of harmony with God) from whence he was found and raised by none other than the Pharaoh's daughter (a symbol of love amidst hate) is simply an allegory stating that while there may be an abundance of darkness and untruth in the hearts of many , the Word of God will never perish but will take root and grow in the hearts of others.

The following story of how "when Moses had grown up" he slew an Egyptian for beating a Hebrew (Exodus 2:11-12), may be taken to symbolize the dark "Egyptian" state of mind succumbing to a more advanced philosophy. That Moses elected to flee from the face of the Pharaoh is to be attributed to the realization that, though higher than the state of mind he had "slain," his spiritual outlook was not yet sufficient to see him through any severe trial. Hence it is recorded that after fleeing he was met at a well by *seven* daughters of the priest of Midian)Exodus 2:16). This meeting of the seven daughters symbolizes the appearance to Moses of insight into the treasured sevenfold process of spiritual advancement, as epitomized by the seven days of creation.

The Episode of the "Burning Bush":

In the third chapter of Exodus we encounter a most interesting example of symbolism. It is recorded that after coming to Horeb, the "mountain of God" (a *mountain* of some sort is frequently used as an expression of *exalted thought*), Moses witnessed the amazing sight of a bush burning with fire, "yet it was not consumed" (Exodus 3:2). Translated into plain language this incident informs us that, regardless of how vivid may be the ravaging effects of imperfect mortal minds (symbolized by the fire), it will never succeed in overcoming or consuming spiritual values (represented by the bush).

The reaction of Moses to this wondrous sight constitutes a rather instructive moral: "And Moses said, I will turn aside and see this great sight, why the bush is not burnt" (Exodus 3:3). If, like Moses, we too could be set thinking and inquiring earnestly into matters of real importance, how quickly would present darkness, fear and superstition give way to the light and comfort of Truth! Only when men stop and attempt to ascertain why (in spite of our own personal prosperity, there happens to be an unquenchable voice within us ever urging us to a sense of universal love) will the world know real peace , prosperity and genuine happiness.

The Three "Signs" Given to Moses:

After the "burning bush" incident, which so inspired Moses to spiritual contemplation, it is reported that the task of spreading the resulting philosophy first appeared too formidable. In answer to his fears and doubts Moses was then given *three* "signs" (see Exodus, chapter 4), or in simple language an increased insight into the powers and nature of God. This deeper perception of reality was to impress upon him some measure of the absolute mastery of Truth— or of identification with God— over such well-known mortal experiences as evil, disease and death. These three "signs" are as follows:

First Sign:

The casting down of a rod (a rod of mortal thought), which "became a serpent," symbolizes the act of exposing our inherently imperfect nature. The subsequent action of Moses following the advice of God (i.e., conscience) and grasping the serpent by the tail, thereby causing it to turn into a rod (in this case a rod of identification with God), signifies that obeying one's conscience will annihilate the false power of evil and endow us with genuine authority and dominion.

Second Sign:

This second sign, involving a leprous hand, concerns faith in the ability of an increased spiritual insight to overcome the injustice of physical disabilities. Just as God may permit physical misfortune, so such misfortune will be seen to be of but a temporary nature, since God is both capable and desirous of overcoming injustice upon a long-term basis.

Third Sign:

The third sign, relating to a pouring of water (a symbol of the Word of God) upon the dry land (a parched and lifeless concept), resulted in the water turning to blood (this time a sign of life) and signifies that acceptance of the Voice of God will inject life into even the most hopeless cases.

The Egyptian Plagues:

Like so many other suspicious events in the Scriptures, the report of Moses supposedly instigating a series of devastating plagues against the Egyptian people (Exodus, chapters 7-9) is not to receive a literal interpretation. Indeed, those who would entertain such a view are actually guilty of the same blasphemy as are holders of the most horrifying doctrine of Original Sin! Why should an entire race be singled out for punishment just because one man (the Pharaoh) refused to "let my people go"? Clearly, no individual of any intelligence could accept a literal interpretation of so obvious an allegory!

Actually, the Pharaoh is used as a symbol of imperfect mortal concepts which tend to hold us in captivity and prevent us from reaching the promised land of Truth. The fact that Moses asked the Pharaoh *seven* times to "let my people go" is most significant, for it symbolizes that the sevenfold process of spiritual advancement was being applied to this false sense of authority. That the Pharaoh required physical proof as to the necessity to let Moses' people go is not at all surprising, since we ourselves can frequently conceive of nothing beyond the physical side of matters. However, in spite of many difficulties, spiritual perseverance must triumph in the end: it is recorded that Pharaoh finally consented to their freedom.

With regard to the various "plagues" themselves, there is every reason to believe that they were used by the prophetic writers to symbolize the inevitable destruction of certain false pagan beliefs which were considered a threat to Hebrew doctrine. Although our modern Bible lists ten plagues, there is evidently confusion as to the nature and number of plagues intended among the early manuscripts from which this story was derived (i.e., the Jahweh, Elohistic and Priestly documents). In any event, taking as an example two plagues which are characteristic of all three documents, we may deduce the following interpretations:

The Nile Turned to Blood:

In early biblical times the Nile was actually worshiped by the inhabitants of Egypt as the very creative center of life— quite the opposite of the true Intelligence we call God, which is the sole source of Life and alone worthy of worship.

The "Slaying" of the First-Born:

This plague, involving the slaying of "all the first-born in the land of Egypt" (Exodus 11:5), refers not to a horrifying slaughter of the first-born of people, but rather of the first-born beliefs held by imperfect man. Since the "first-born" beliefs of mankind can only be described as "Egyptian" or pagan in nature, it will be seen that in order to advance spiritually, it is most essential that our own "first-born" be slain also.

Although having touched but lightly upon the wealth of ideas and morals symbolized by such episodes, it should nevertheless be possible for one to go ahead on personal initiative and apply these same basic principles of symbolism to other allegorical writings. For instance, continuing the saga of Moses, it will be found that the celebration of the Passover, the *seven* ascents of Mount Sinai (in which he received *Ten* Commandments) and even the instructions regarding the building of the Tabernacle all convey messages of a spiritual nature. Moreover, the statement that Moses was 120 years old when he died (Deuteronomy 34:7) is also quite interesting, inasmuch as the symbols *one, ten* and the sense of the *second* day of creation may be discerned. Interpreted, it signifies that Moses recognized one God and sought to impress this principle upon his people, and that at the time of his death he had, in some measure, succeeded in separating good from evil— that he had led them out of the land of "Egypt," or darkest thought.

So-called "Miracles":

There was once a time when all the reported "miracles" of the Bible were accepted without question as being authentic historical events. But this is an age of rapidly dawning consciousness in which such views must submit to reason. In fact, the very nature of any miracle— or divine intervention— strikes at the ethical structure of the universe; for it involves what is nothing short of *favoritism!* Why should any particular person (or persons) be singled out, in a truly special manner, to experience some sort of extraordinary experience which is against the very laws of nature? Is it not clear that an outstanding feature of the cosmos lies in its consistency— that in order to be just it cannot possibly contradict itself?

The Crossing of the Red Sea:

Take, for example, the story of the "miraculous" crossing of the Red Sea as described in Exodus, chapter 14. To assume that God would intervene directly and in such dramatic fashion to save one nation from its enemies, without showing the same favor to others faced with comparable predicaments, is quite contrary to all sane reasoning. God most definitely plays no favorites! In reality, the reported crossing of the Red Sea does not refer to a mere physical passage through a physical medium, but rather symbolizes a *spiritual passage through a spiritual medium!* It symbolizes the passage of man from the darkness and fear of the "Egyptian" state of thought to a superior and more concrete state of mind (i.e., the "dry land" state of the third day of creation! As a matter of fact, there are three distinct stages of thought involved in this allegory which correspond perfectly to the principles expounded in the first three days of creation:

Day One:

"Then Moses stretched out his hand over the sea, and the Lord drove the sea back by a strong east wind all night" (Exodus 14:21) ... "and the Spirit of God was moving over the face of the waters" (Genesis 1:6).

Day Two:

"And the waters were divided" (Exodus 14:21) ... "Let there be a firmament in the midst of the waters, and let it separate the waters from the waters" (Genesis 1:6).

Day Three:

"And the people of Israel went into the midst of the sea on dry ground" (Exodus 14:22) ... "and let the dry land appear" (Genesis 1:9).

During the course of our earthly lives we are all faced with the challenge of having to pass over some similar obstacle— of having to improve and expand our perspective in order to escape the effects of materialism and evil. Applied to everyday experiences, the overall moral of this instructive allegory may be stated as follows: Whenever confronted with fears and doubts, and with difficulties which seem to us to be insurmountable, if we will but press onward, ever trusting in the light of conscience to guide us, we shall soon find ourselves to be situated upon dry land, our fears and doubts will eventually be washed away (just as the :Egyptian" state of mind was "drowned"), and we will have succeeded in crossing our own "Red Sea."

The Crossing of the Jordan:

The story of Joshua leading the Israelites across the River Jordan is to be considered a sort of sequel to the earlier crossing of the Red Sea and symbolizes a further advance in spiritual understanding. However, unlike the previous crossing, which was made of necessity and tinged with an element of fear, the crossing of the Jordan is characterized by a sense of *authority* stemming from a superior outlook (the *twelve* men carrying the *twelve "stones"*). This increased spiritual vision is well illustrated in the statement that *seven* tribes (symbolizing a false or opposite sense of the seven days of creation) would be driven out before them (Joshua 3:10). The fact that the ark of the covenant was carried before them signifies exalted thought and an earnest search for Truth. Also of great interest is the statement that "The people came up out of the Jordan on the *tenth* day of the *first* month" (Joshua 4:19), as it symbolizes the recognition of One Supreme Cause (or God) and of the duty to apply such insight to matters of human experience.

The Falling of the Walls of Jericho:

Here we come to an allegory permeated so conspicuously with the sacred or symbolic numerals (particularly the number "seven") that it is positively amazing how anyone could read it without being forced to abandon all notions of a literal interpretation. First of all, it is significant to note that *"forty thousand* ready armed for war passed over before the Lord for battle, to the plains of Jericho" (Joshua 4:13). Invariably, the authors of this story were attempting to stress their quest for Truth, and to put their presumably superior knowledge into practice, The statement that *"seven* priests shall bear *seven* trumpets of rams' horns before the ark, and on the *seventh* day you shall march around the city *seven* times, the priests blowing the trumpets" (Joshua 6:4) signifies application of the sevenfold process of spiritual advancement. The subsequent fall of the walls of Jericho does not constitute a miracle involving mere physical walls at all, but rather symbolizes the fall of much more important "walls of ignorance" which would bind us to a *false sense of reality!*

The Story of Jonah:

The impressionable story of Jonah and the whale is clearly one more instance of the prophetic writers resorting to fiction in order to express a moral or principle. Essentially, the message imparted by this narrative stresses the obligation to follow conscience, even if it should be less than enlightened.

The city of Nineveh represents those individuals who, while they would readily accept the Truth if they were aware of it, nevertheless live in a state of ignorance. (In Jonah 4:11, it is reported that they "do not know their right hand from their left.") Jonah, aware of this unfavorable prospect, disobeyed the command to preach to the people of Nineveh and instead attempted to get away from conscience: he "rose to flee to Tarshish" (Jonah 1:3). Cast into a stormy sea, he is supposedly swallowed by a "great fish." After *three* days and *three* nights in the belly of the fish he is rather magically deposited on dry land. Given time and incentive to reconsider matters, Jonah's conscience finally wins out and he repents his action after acquiring a higher perspective (i.e., the "three" aspect of his ordeal).

With renewed vision, Jonah then faced the duty of preaching in Nineveh, which was reported to be a *three-day* journey. The fact that the people of this city were forgiven when they repented of their misdeeds, even though many were still incapable of distinguishing between right and wrong, imparts the moral that sin must really be defined as a personal disobedience of conscience— quite irrespective of caliber.

Summary

While we have presented a brief insight into some of the more popular characters and events of the Old Testament, of necessity many others of interest have been omitted.

For instance, in I Samuel, chapter 17 there is the colorful and inspiring story of David and Goliath, in which the latter is depicted as a symbol of materialism. (It is stated that the height of Goliath was "*six* cubits and a span," that the weight of his coat of mail was "*five thousand* shekels of brass," and that his "spear's head weighed *six hundred* shekels of iron.") That Goliath (i.e., materialism) was slain or overcome by a *stone* wielded by David is not surprising, since a "stone" is often used as a symbol of the *appearance of Truth*, which will slay any sense of evil.

Then again in II Kings (chapter 1) there is the episode of Elijah commanding fire to come down from heaven. The symbolic nature of this allegory should be obvious even to one not versed in the system of biblical symbolism; for to believe that God would intervene in such dramatic fashion to destroy *one hundred* men who could not be deemed any worse than the average, is quite unthinkable! Actually, the instructive moral behind this story, which centers around a reported persecution of Elijah because he dared to criticize King Ahaziah for his lack of monotheistic faith, is simply that those who fail to recognize one God will surely suffer as a consequence. For example, the first two attempts to take Elijah resulted in the "death" of two groups of *fifty* men (which symbolizes the opposite of the fifth day of creation— *exalted thought)*; it was only after a *third* attempt (when insight into the Truth was added) that their lives were secure.

In spite of a somewhat restricted analysis of the Old Testament, it will nevertheless be seen how futile— and foolish— it would be to attempt to reduce the various passages to mere physical events. (Some characters and incidents are obviously entirely symbolic; with others it is difficult to tell if some literal consideration might not be justified.) However, once it is realized that it is only *Principle* which is of fundamental and lasting significance, it follows that the question of whether a given narrative is fact or pure fiction is really of minor concern. What is important is that sentiment must not be allowed to prevail over the revelations of science and plain logic.

9

The New Testament

The Enigma of Satan— The Jesus/Christ Paradox— Jesus: His Birth and Mission— Spiritual Significance of the "Miracles"—The Crucifixion and Resurrection— The "End of the World" Discourse— The Book of Revelation— Words of Lasting Wisdom

Although of a generally higher nature than many documents of the Old Testament, and while dealing with the life and teachings of a most dynamic personage, it remains that the New Testament is primarily a *continuation of earlier symbolic and spiritually significant writings!* To not a few earnest seekers of Truth, it is bound to come as somewhat of a shock to learn that traditional interpretations are in need of considerable revision; but this is certainly the case. Indeed, even such basic titles as "Satan" and "Christ" demand an explanation which is far different from that offered by orthodox theology. In this age of science and enlightenment we must disregard outworn concepts— concepts which are the product of mortal men with wild imaginations. It is clearly time to open our minds to common sense! [1]

The Enigma of Satan

Throughout the ages the identity of Satan has proven to be a source of great embarrassment to many theologians. For while they have been inclined to look upon the devil as some sort of supernatural being and opponent of God, they have been quite unable to explain a rather fundamental mystery concerning his very existence. In fact, such theology is guilty of holding two highly conflicting points of view; for if God is really omnipotent, *then why does He simply not destroy Satan?* Obviously, any doctrine which would consider Satan to be a spiritual entity of a supernatural status is quite ridiculous; and the very prevalence of such a belief may once more be traced to a distorted biblical interpretation.

Actually, the devil is nothing more than a particular *symbol* that was introduced into the Scriptures for the purpose of illustrating a *lack of Truth!* The entire problem of evil is but a false sense of reality. Indeed, is the symbolism portraying lack of the Truth not perfectly clear when we recall Jesus' reference to the devil? "He was a murderer from the beginning, and has nothing to do with the truth, because there is no truth in him. When he lies, he speaks according to his own nature, for he is a liar and the father of lies." (John 8:44).

What we might like to consider so conveniently as being the works of the devil should really be attributed to the results of our own selfish desires— our own state of imperfection! Whether or not it is realized, all sin originates from an unbalanced love of oneself; *for selfishness is at the root of all evil!* As we grow older, it is true, it may manifest itself in many different ways, but nevertheless all of the evil thoughts which may materialize in our minds can be traced to this one source. This being so, then we must certainly admit that our self-centered minds possess all the cunning and deceitfulness that the Scriptures would appear to ascribe to Satan. For it is none other than our own ignorant and selfish state which is ever working, often with great deception, to mislead us and blind us to the Truth.

If we would attempt to lead a sinless life, we must be made to understand that Satan is simply *our own inherently selfish nature*; for once we do, it will enable us to meet temptation on more equal terms. Whenever temptation arises (no matter in what form), we should make an honest attempt to ascertain if our own selfish interests might not play a far greater role than we would like to believe. If only we would do this, then the chances of deceiving ourselves would be much reduced. We will have succeeded in bringing temptation out into the open, where it can then be seen in its true perspective.

The Jesus/Christ Paradox

Our fundamental problem is one which orthodox Christianity as a whole claims to understand; but when investigated, its extremely dogmatic explanation leads only to contradiction and nebulosity. In short, who (or what) is a "Christ?" What is the true status of "Jesus," that historical personage whom so many hold in such high esteem?

On the basis of orthodox views, we are apparently to believe that Jesus was a one and only "Son of God"—that no other comparable entities ever existed. This notion can only be described as highly naive, for even if we were to accept the traditional idea that Jesus constituted our world's one and only "Savior," it sadly fails to take into consideration all the countless numbers of *other worlds* similar to our own which have existed in the past, exist now and which are bound to exist in the future! Surely we cannot be so vain as to deny these other worlds the light of a "Christ" such as ours— or do we have so much as the slightest idea as to the true nature of the spiritual cosmos? Clearly, this popular concept of a "Savior" is not a very realistic one and breaks down completely under minimal reasoning.

Upon examination of the Bible itself, we encounter a rather confusing and contradictory state of affairs. On the one hand, we have the statement "I and the Father are one" (John 10:30), with the curious doctrine of a so-called Holy Trinity based upon the belief that Jesus was actually *equal* to God; while on the other hand we have numerous passages which impress upon us the understanding that God is *greater* than Jesus: "the Father is greater than I" (John 14:28); "not my will, but thine will be done" (Luke 22:42); "I do nothing on my own authority but speak thus as the Father taught me" (John 8:28); "If you keep my commandments, you will abide in my love, just as I have kept my Father's commandments and abide in his love" (John 15:10); "And Jesus said to him, 'Why do you call me good? No one is good but God alone.'" (Luke 18:19).

The true solution to this glaring contradiction (which orthodox theology has yet to explain) is simple in essence. Recalling that the Scriptures deal primarily with principles, and that historical events are of secondary importance, it should not come as a surprise to learn that Jesus was described initially— and *symbolically*— in terms of a "dual" personality. That is to say, when referring to Jesus' intrinsic human status it must certainly be acknowledged that "the Father is greater than I," since we are dealing with an entity characteristic of mankind— a "son of man." In contrast to this imperfect state of mind we have the "Christ" or "Son of God" state, which is really a *title or goal* in that it ***symbolizes*** attainment of that state of Perfection that is God Himself: "I and the Father are one." Thus it will be seen that, throughout the course of his earthly ministry, Jesus was striving to demonstrate and live the ***hypothetical*** goal or ultimate "Christ" state of Perfection. This principle of attempting to reach such a priceless goal we may term the **"Christ-idea"**— for essentially, "Christ" is a *spiritual idea or ambition* of which it may well be said that "I am the way, and the truth, and the life; no one comes to the Father but by me." (John 14:6).

It will, therefore, be concluded that Jesus himself was a very human being, and all attempts to picture him as anything else will be seen to be rooted in sheer fantasy!

Jesus: His Birth and Mission

As one might suspect, the attempts of Matthew and Luke to trace the genealogy of Jesus back through the ages are not to receive a literal interpretation. On the contrary, they serve to convey two messages of a distinctly spiritual nature. Is it not a rather strange coincidence to note that in the genealogy listed by Luke, *Jesus happens to be the 77th name?* Recalling the symbolic meaning of the numeral "seven," the reason for this is obvious: it symbolizes that Jesus was "fulfillment"— that he was indeed capable of making an impressive demonstration of the *Christ-idea.* In the genealogy that is recorded in Matthew, is it not equally significant that there are *three* very distinct groups, each of which contains *fourteen* ("ten" plus "four") generations (Matthew 1:17)? In other words, Matthew is conveying the message that, in the course of his endeavor to seek the Truth (i.e., the "four"), and apply such light to human experience (i.e., the "ten"), Jesus would inevitably give a famished world welcome insight into the nature of the spiritual cosmos (i.e., the "three" aspect).

Incidentally, it must be clear that efforts to trace the genealogy of anyone back thousands of years at that early date can only be regarded as absurd; for even today it is practically impossible to track our own ancestry back more than a few hundred years. So we see that all things considered, there is really no reason to suppose that the authors of such genealogies actually believed in the historical truth of their stories, although it would no doubt suit their purpose quite well if the masses could be persuaded to give them credence.

The recorded birth of the babe Jesus contains many examples of symbolism, of which the following instances will be found to convey instructive morals:

"Behold, a virgin shall conceive and bear a son." (Matthew 1:2)

The real meaning behind this "virgin birth" story involves our recognition that the "Christ-idea" (i.e., the principle of rising to a state of Perfection) will only be born to one possessing "virgin thought"—to one who has not been contaminated with false theology.

"And she gave birth to her first-born son and wrapped him in swaddling clothes, and laid him in a manger; because there was no place for them in the inn." (Luke 2:7)

Here we have a beautiful example of symbolism. The birth in a "manger" symbolizes that there is never any room (or time) for the "Christ-idea" to enter into the hearts of well-to-do people concerned with worldly affairs (i.e., the "inn" state of mind); for it can only be born to those of a humble or lowly "manger" state of mind.

"Now when Jesus was born in Bethlehem of Judea in the days of Herod the king, behold, wise men from the East came to Jerusalem, saying, 'Where is he who has been born king of the Jews? For we have seen his star in the East, and have come to worship him." (Matthew 2:1-2)

This "star in the East" clearly represents the guiding light of conscience, which any wise man will follow.

"Then Herod summoned the wise men secretly and ascertained from them what time the star appeared." (Matthew 2:7)

Herod may well be seen to depict that mortal state of mind which always wants to take things of a higher spiritual nature and reduce them to material terms concerned with such trivialities as time, persons and places.

"Then, opening their treasures, they offered him gifts, gold and frankincense and myrrh." (Matthew 2:11)

This story of the wise men presenting *three* gifts to the infant Jesus may be interpreted to mean that he would indeed be given insight into the Truth.

"Now when they had departed, behold, an angel of the Lord appeared to Joseph in a dream and said, 'Rise and take the child and his mother, and flee to Egypt." (Matthew 12:13)

The fact that it was imperative to flee from Herod for a time symbolizes that, in the face of evil, it is sometimes advisable to take the newborn spiritual idea and hide its light until sufficiently developed to assure complete triumph.

"Then Herod ... sent and killed all the male children in Bethlehem and in all the region who were two years old and under." (Matthew 2:16)

This "slaying" of young children is similar to that reported at the time of the birth of Moses; it symbolizes that the world is forever seeking to destroy those who would want to separate good from evil.

From the recording of the birth of Jesus until the actual start of his mission, we encounter some interesting examples of symbolism. First of all, it will be noted that the only mentioned incident of the childhood of Jesus is at the age of *twelve*, where he was lost for some *three* days and eventually found "in the temple, sitting among the teachers, listening to them and asking them questions" (Luke 2:46). Recalling the symbolic meaning of the numeral "twelve," the actual circumstances under which he was found the "third" day, and Jesus' explanation that "I must be in my Father's house," we arrive at the following interpretation: In order to demonstrate the universal scope and authority of divine Principles to others, we must also be "three" days in our Father's house (i.e., we must first obtain insight into the Truth.

After this episode at the age of twelve years we next hear about Jesus at the age of *thirty* years ("three times "ten"), when he is reportedly baptized preparatory to setting out upon his preaching mission. Whether Jesus really started his ministry at the age of thirty years (Luke 3:23) is quite unimportant; the message intended is that Jesus possessed insight into the nature of spiritual matters (i.e., the "three"), and was desirous of passing this knowledge on to others (i.e., the "ten").

The very act of baptism is one which is obviously misunderstood by theologian and layman alike. Introduced into biblical literature as a *symbol* of imparting to one some measure of insight into the Truth, the idea has since gotten completely out of hand and become little more than an empty ritual. We now witness priests going through the motion of sprinkling actual water upon individuals in the vain belief that their efforts are able to raise such souls in the esteem of God. As a matter of fact, not only are adults subjected to a pointless physical ritual, but the general intention seems to be to baptize tiny infants quite incapable of thought! Clearly, a mere symbol has become distorted beyond all reason. Only in the sense that a person has received a more realistic insight into the Truth may it be said that one has been truly baptized; all else is to no avail.

After his baptism, signifying that he indeed possessed insight into Truth, Jesus then experiences *three* temptations by the devil. Here we have an instance in which the symbol "three" is used in its opposite sense; for it signifies that our own imperfect nature (i.e., the "devil") is forever tempting us with a false perspective. From the report that Jesus "fasted *forty* days and *forty* nights" in the wilderness (Matthew 4:2) we may conclude that he was earnest in his desire to seek and live the Truth. (The term "wilderness" is used throughout the Scriptures as a place where the material sense of things gives way to the spiritual.)

Coming to the actual mission itself, it is most interesting to note that Jesus chose *twelve* apostles— a number symbolizing the authority and universal extent of divine Principles. An investigation of the calling of the apostles also reveals much of a symbolic nature. For instance, in Matthew (4:18-22), is it not instructive to read that Jesus called *four* followers initially (Peter, Andrew, James and John); while in Luke (5:1-11) there are also *four* mentioned on a "fishing" expedition (Jesus, Peter, James and John)? Can we not discern here the same symbolic message as in the saga of Abraham where *four* people are listed as setting out for the "land of Canaan" (i.e., in search of Truth)? What we need today is more individuals who would make it a priority to become "fishers of men."

Spiritual Significance of the "Miracles"

Invariably, the so-called "miracles," which center around the life of Jesus, are just as thoroughly embedded in allegorical structure as the many earlier instances of the Old Testament. (Is it not perfectly clear that the very idea of a "miracle" is against Cosmic Law?) Upon applying our knowledge of biblical symbolism to certain of these purported events we may derive the following interpretations:

The Feeding of the Multitudes:

As misguided theologians of the past first saw fit to teach, so many of our present-day churches exhibit little improvement and still tend to look upon Jesus' "feeding of the multitudes" in a mere earthly and materialistic sense. Taken literally, such an incident is not only incomprehensible, but is quite contradictory with respect to an earlier occasion where Jesus refused to turn a stone into bread that physical wants might be satisfied. On the contrary, far from feeding the multitudes with physical food, Jesus fed them with *spiritual* food!

Can we not see that those instances of dispensing "bread" refer not to relatively unimportant physical food, but to the *Word or Bread of Life?* "Jesus said to them, "I am the bread of life; he who comes to me shall not hunger and he who believes in me shall never thirst." (John 6:35). "This is the bread which came down from heaven, not such as the fathers ate and died; he who eats this bread will live forever." (John 6:58). Is not the symbolic aspect of the "bread" quite unmistakable when we further read Jesus' warning not to eat of the leaven of the Pharisees and of the Sadducees? "How is it that you fail to perceive that I did not speak about bread? Beware of the leaven of the Pharisees and Sadducees. Then they understood that he did not tell them to beware of the leaven of bread, but of the *teaching* of the Pharisees and Sadducees." (Matthew 16:11-12).

Upon looking deeper into the two occasions where Jesus is reported to have fed the multitudes, the whole affair becomes crystal clear once we apply our knowledge of the symbolic numerals. For instance, in Matthew, chapter 14, it is recorded that *five* loaves and *two* fishes were divided among *five thousand*, from which *twelve* baskets of food were later gathered. As the "five" is used here in the sense of the fifth day of creation (i.e., fullness of Life) and the "two" in the sense of the second day (i.e., separation of good from evil), it will be seen that when such a philosophy has been digested by those desirous of obtaining a higher and more abundant sense of Life (i.e., the "five thousand"), it will result in the Word being turned into a successful demonstration of the power and universal extent of divine

Principles (i.e., the "twelve" baskets of fragments).

In Matthew, chapter 15, we have a slightly different rendition. In this case it is said that *seven* loaves and a few fishes were distributed among *four thousand* men who had fasted *three* days after ascending high ground (symbolizing exalted thought) and that *seven* baskets of food were left over. Reduced to simple language, we have the message that Jesus imparted to those who were really desirous of seeking the Truth (the "four thousand" who had fasted "three" days) an abundant sense of Life (the "fishes") and insight into the spiritual meaning behind the seven days of creation (the "seven" loaves) which, if given time and utilized properly, are bound to lead to multiplication, fulfillment and Perfection (the "seven" baskets).

The Water Turned to Wine:

This traditional "miracle" (John 2:1-10) is obviously another allegory used to illustrate a principle of lasting value. The overall moral to be drawn from this story may be stated as follows: Whenever the Word of Life (symbolized by the *water*) enters into a false sense of reality (epitomized by the *six* stone jars), it will effect a change and produce *inspiration* revealing that Love is supreme (*wine* is used throughout the Scriptures as a symbol of the inspiration of Love). In other words, to the extent that we let the *Christ-idea* enter our hearts it will inspire us and lift our souls to a truer and more filling sense of Love.

The described marriage feast does not, of course, refer to the usual union of two members of the opposite sex, but rather to the "wedding" of human minds to a higher and more advanced outlook upon life. Hence it is reported that "On the *third* day there was a marriage at Cana in Galilee" (John 2:1); the participants indeed received an insight into the nature of spiritual matters.

The Man Healed of the "Withered Hand":

This particular episode, in which a man's "withered hand" is alleged to have been restored physically, must likewise submit to reason.

Does it not strike one as rather odd that the Pharisees, before whom this act was performed, failed to be impressed by such an incredible event? Would one not expect even the most unenlightened priest to immediately bow down in sheer fright and worship the performer of so vivid a miracle? Of course one would! But what was their reaction? They actually "went out and took council against him, how to destroy him" (Matthew 12:14).

In essence, the moral which the author of this story is attempting to convey is that the false and enslaving theology of the Pharisees, which would reduce an individual's duty to seek the Truth to a slothful reliance upon the views of so-called "authorities," had withered and deadened the people's sense of reality. We can readily see why the Pharisees were not impressed by any "physical" miracle but sought only to destroy Jesus, since his teachings were succeeding in opening the public's eyes to the necessity of *thinking for themselves* and to the utter futility of relying upon their withered philosophy. If only the world could free itself from the pernicious grip of its modern Pharisees!

The Raising of Lazarus:

Like other biblical accounts involving the restoration of sight and physical health, the raising of Lazarus is an allegory which illustrates a definite moral. To mortal eyes a man appears dead if he is inert physically; while in reality the only true death is one of a spiritual nature. Hence, although mortal man might consider one to be dead in the physical sense, to another with a higher and more advanced outlook it will be realized that the spiritual (i.e., one's soul) is never dead— being the only true reality and the only intrinsic source of Life.

Examining the story of the raising of Lazarus, we promptly come across a most interesting discovery; for he is reported to have "been in the tomb *four* days" (John 11:17). Clearly, the moral behind this particular example of symbolism is that, regardless of mere physical occurrences, if one has been earnest in his quest for Truth he cannot be dead, since the spiritual is the only true source of Life and the only real death. Consequently, it will be seen that the *stone* which was taken away (John 11:41) refers not to any material stone, but to the removal of that false perspective which would consider physical death to be all-important (the "stone" is used here in its opposite sense). How much more serious is it to incur spiritual death, in preference to a temporary physical death!

Can we not see that such instances, which orthodox theology would take in an earthly physical sense, really refer to the far more important spiritual sense? Is it not clear that, throughout the course of his ministry, Jesus instilled into people a more realistic sense of Life and brought many who were sick spiritually (or even "dead" to the Truth) to real Life and spiritual health? No wonder it is reported that as many as touched the hem of his garment "were made well" (Matthew 14:36), since whenever one touches or comes to accept the *Christ-idea*, any spiritual sickness (or "blindness") will very soon become past history.

The Walking on the Sea:

Among the traditional "miracles" are those of Jesus walking on the sea and the stilling of the tempest. However, just like so many other distorted theological concepts which erroneously neglect the spiritual in favor of the physical, a drastic revision is again indicated. For instance, the "sea" incident is not to be taken as referring to any physical body of water at all; but rather the term "sea" is used over and over again in the Scriptures to symbolize exposure to *adverse and worldly influences* Accordingly, it will be seen why Jesus was so successful in stilling the "tempest," since error and a materialistic mode of thought will naturally vanish before the virtues of a superior philosophy.

In connection with the episode of his walking on the sea, is it not instructive to note that Jesus came walking to his disciples "in the *fourth* watch of the night" (Matthew 14:25)? Actually, the reason for the disciples finding themselves "beaten by the waves, for the wind was against them" (Matthew 14:24) was that they had not gone with

Jesus up into the mountain to pray. That is to say, they had not exerted the necessary thought to grasp or digest the full meaning of the "spiritual feast" which had just taken place. Nevertheless, in spite of periods of temporary regression, if in our hearts we really desire to draw closer to the Truth we will not be denied (i.e., Jesus, representing the *Christ-idea*, is reported to have drawn toward the disciples in the *fourth* watch). The subsequent incident, in which Peter is said to have attempted— unsuccessfully— to duplicate Jesus' feat of walking on the sea, conveys the moral that one must have faith in the *Christ-idea* and not succumb to the ravaging effects of an imperfect and hostile world.

The Restoration of the Severed Ear:

The incident in the garden of Gethsemane , in which the slave of the high priest is first supposed to have his right ear cut off by one of the apostles and then to have it so miraculously restored by Jesus (Luke 22:50-51), is so obviously meant to be interpreted symbolically that it is puzzling why any theologian would ever think otherwise. Can anyone honestly believe for one moment that these soldiers could have witnessed so spectacular a miracle and then promptly arrest Jesus? Hardly likely! Much rather would they have quaked with fear of what they were about to do.

In reality, the instructive moral behind this story may be stated as follows: Do not be a foolish zealot who, like Peter, would in one instance pass up a golden opportunity to gain insight into the nature of God ("Could you not watch *one* hour?) and then when a crisis arises would prefer to rely upon mere physical powers and "cut off somebody's ear." An understanding of the real nature of the spiritual cosmos is a far greater asset to one than any physical superiority; and in the face of temptation to resort to physical means in order to solve a problem we would do well to remember this very point: "Then Jesus said to him, 'Put your sword back into its place; for all who take the sword will perish by the sword." (Matthew 26:52).

The Blinding of Paul for Three Days:

The story of Paul being "blinded" for *three* days, while on the road to Damascus, has unfortunately been the subject of another literal interpretation by misguided theologians who apparently can see nothing wrong with a God who "selectively" interferes with the very laws of nature. The truth about this narrative, of course, is that it was written to symbolize the dawning of the light (i.e., the *Christ-idea)* to Saul, who was subsequently renamed Paul, just as certain other biblical characters were similarly renamed in order to signify their new relationship with God. The "light from Heaven" which "flashed about him" (Acts 9:3) refers to no mere physical light, but to the *enlightening revelations of conscience*, which indeed inquired of Saul: "Saul, Saul, why do you persecute me?"

After Paul was *three* days "without sight, and neither ate nor drank" (Acts 9:9), it is recorded that "immediately something like scales fell from his eyes and he regained his sight" (Acts 9:18). In other words, Paul had passed through a state of mental conflict only to emerge triumphant and in the possession of a much higher outlook. Whatever the eventual consequence, the conversion of Paul stands as a remarkable example of repentance and noble sacrifice for the sake of conscience.

The Crucifixion and Resurrection

Like other biblical narratives written primarily to illustrate a moral or principle, the events surrounding the crucifixion and alleged resurrection on the *third* day are no exception. For while they may indeed be based upon the actual death of Jesus, the use of symbolism is nevertheless quite pronounced. A brief insight into a few examples of this may be gained from the following:

The Last Supper:

As unfortunately proved to be the case with regard to baptism, so theology has once again seized upon symbolic writings and concocted yet another senseless ritual. First of all, the "large upper room furnished" which a man "carrying a jar of water" showed Peter and John (Luke 22:10-12) refers not to any mere physical event, but rather symbolizes that the "water" (i.e., the *Word of Life*) will lead one to a state of exalted thought— to a superior outlook as expressed by the "large upper room furnished." The last supper itself, at which Jesus is reported to have distributed "bread" and "wine" to his apostles, is not to be taken literally as some would have us believe. Instead, it was a distinct sense of Life and the inspiration of Love which Jesus imparted to his disciples. Hence the words "this do in remembrance of me" really refers to "do likewise" and impart to others a higher sense of Life and Love. To believe that the popular physical ritual of a so-called Holy Communion is capable of imparting to one anything of value, as a result of the presence of actual bread and wine, is quite ridiculous. It is strongly opposed to both common sense and a stable cosmos, and has nothing at all to do with genuine religion.

Judas Betraying the Christ-idea:

Taken literally, the Judas episode would convey the horrifying impression that there is such a thing as predestination. (To believe that our very actions are predetermined is to deny the principle of free will!) Actually, Judas is portrayed as a symbol of imperfect mortal thought, which is ever bound to betray or renounce the Truth (i.e., the *Christ-idea*). Is it not instructive to note that Judas betrayed "Christ" for *thirty* pieces of silver, symbolizing exactly the opposite of putting spiritual insight into practice? How often are we ourselves guilty of betraying the *Christ-idea* for some mere material and transitory advantage!

The Three Denials by Peter:

As a sort of sequel to the moral of Judas betraying "Christ" for *thirty* pieces of silver, we have the *three* denials by Peter: "Jesus said to him, 'Truly, I say to you, this very night, before the cock crows, you will deny me *three* times.'" (Matthew 26:34). In this case the "three" is also used in its reverse sense, signifying a spiritual lapse or regression. The fact that three denials were to take place before morning symbolizes that before the light finally dawns upon one there will be occasions when the *Christ-idea* will be rejected and forgotten. Clearly, we have a great contrast between Peter's three denials and the moving drama in the garden of Gethsemane, where Jesus prayed *three* times that the "cup" might pass from him. (This "cup" represents the agony of seeing the spiritual ideas which he so highly cherished be misunderstood, ignored and defiled by mortal minds.)

The Releasing of Barabbas:

Barabbas is introduced into the Scriptures as a symbol of highly imperfect mortal thought. Hence the choosing of Barabbas over the "Christ" is not at all surprising, for even today in this present age the world still seems to prefer the "Barabbas" state of mind to an ideal which requires sacrifice and the expenditure of much effort. How frequently have we all, at one time or another, been guilty of choosing the Barabbas state in preference to the far more demanding *Christ-idea!*

The Crucifixion:

As a sample of the rich symbolism, which permeates the biblical record of the crucifixion, we may cite the following examples:

"And it was the third hour, when they crucified him." (Mark 15:25)

It was certainly an hour devoid of spiritual vision (the "three" is used here in the opposite or negative sense)!

"And with him they crucified two robbers, one on his right hand and one on his left." (Mark 15:27)

The two robbers crucified with Jesus (making *three* in all) may be taken to signify what the world is ever threatening to do to one with a superior outlook. Also, the two robbers themselves signify that the world is frequently unable to discern right from wrong.

"When the soldiers had crucified Jesus they took his garments and made four parts, one for each soldier." (John 19:23)

Jesus' garments may be taken to represent divine Principle, and the action of dividing his garments into *four* parts signifies that they "split up" and therefore failed to recognize the Truth (the "four" being used in its reverse sense to convey failure to seek the Truth).

"But his tunic was without seam, woven from top to bottom, so they said to one another, 'Let us not tear it, but cast lots for it to see whose it shall be.'" (John 19:23-24)

The fact that they "cast lots" for his tunic "without seam" imparts the moral that many do not have a genuine desire to work for a higher perspective and so are willing to "take a chance" on whether or not the undivided Truth will come to them of its own accord. (Invariably, many individuals find themselves with a false sense of values simply because of reluctance to expend effort on behalf of a superior philosophy, choosing instead to follow the easiest course and accept whatever views happened to be the most popular.)

"And he said to him, 'Truly I say to you, today you will be with me in Paradise." (Luke 23:43)

This reply by Jesus to the repentant thief on the cross expresses the moral that God is of a forgiving nature and that what will determine a person's future status, quite regardless of past deeds, is the actual state of one's soul at the time of death.

"It was now about the sixth hour, and there was darkness over the whole land until the ninth hour, while the sun's light failed; and the curtain of the temple was torn in two." (Luke 23:44-45)

There is indeed darkness at any "sixth" hour ("six" is often used as a symbol of mortal thought) which will last until the "three" (i.e., spiritual insight) has been added, making the "ninth" hour. With possession of genuine knowledge the "curtain of the temple" (a symbol of false and blinding theology) will in truth be "torn in two"; it will be utterly destroyed.

The Rising on the Third Day:

Clearly, what is meant by this rising on the *third* day is that the righteous nature of Jesus must assure a *spiritual resurrection.* What invariably took place was an instantaneous resurrection (time does not exist as such between incarnations) in which the soul of Jesus was permitted the higher manifestation so richly deserved. (In this instance the "three" symbolizes insight into the principle of spiritual immortality.) That this rising on the "third" day was to be considered symbolic is borne out most conclusively by the statement: "For as Jonah was *three* days and *three* nights in the belly of the whale, so will the Son of man be *three* days and *three* nights in the heart of the earth" (Matthew 12:40), *since the episode of Jonah and the whale is so obviously an allegory!*

Upon looking over the various stories concerning discovery of the "resurrection," we note that emphasis is placed upon the "stone" which was "rolled away from the tomb." The tomb represents imperfect (or "dead") human hearts; the *stone* is this time used in the reverse sense to signify the manifestation of a false philosophy. Hence, on the *third* day the appearance of spiritual insight will have served to roll away a false perspective from human hearts. The result of this rolling away of untruth was the realization that "He is not here; for he has risen" (Matthew 28:6). For the "Christ" is not a mere physical body tangible to physical senses; but rather the *Christ-idea* depicts a *spiritual goal*— of the process of rising to a state of Perfection— and is therefore capable of being perceived only by those inclined spiritually. The "angels" in white and "shining garments," which are reported to have informed certain of Jesus' followers that he had risen, are introduced as "messengers of God" and may be taken to depict the revelations of conscience.

Three Appearances of the "Christ":

As one might suspect, the *three* appearances of the "Christ" are really attempts to express a moral or principle. Essentially, they are intended to signify that the apostles of Jesus would succeed in their quest for Truth (symbolized by the *three* appearances of the "Christ" unto them.

Taking the Gospel of John as an example, these three alleged appearances may be contemplated as follows:

(1) *The appearance to Mary Magdalene:* It is interesting that the initial reappearance of "Christ" (i.e., of the *Christ-idea*) should be made at the tomb where, moments before, there was nothing to see. Mortal man is bound to search for things of a physical nature. It is only after light has been added and Truth has been separated from falsehood (epitomized by the two "angels in white") that one is able to grasp that "Christ" is really an *Ideal or Principle*, which no man can destroy and which must ever be with us even though we might have no physical or visual proof. Physical senses can visualize no more than material phenomena; while conscience is able to perceive things of a much higher spiritual nature.

(2) *The appearances to the apostles in a closed room:* The two "closed door" appearances, once with Thomas absent and once with him present some *eight* days later— signifying a search for Truth and success in separating good from evil (i.e., the "eight" or "four" times "two" aspect)— conveys two morals. First of all, the "closed room" symbolizes the same open-minded state of thought as does the "virgin birth" moral; for once we close our doors to all adverse external influences we must leave our hearts open for the appearance of the *Christ-idea* (i.e., the revelations of conscience). Secondly, the "doubting Thomas" episode expresses the moral that it is wiser to believe in principles revealed by conscience than to accept only what is perceived by our imperfect and limited physical senses.

(3) *The appearance to the apostles who were fishing:* The fact that the apostles could not discern the *Christ-idea* until they had first cast their nets on the "right side of the boat" (John 21:6), may be taken to mean that one's thinking will bear no fruit unless such virtue is utilized correctly. The actual catch itself, "a hundred and fifty-three" large fish, is interesting inasmuch as it contains the basic symbols "one" and

"ten" along with the "three" and the sense of the "fifth" day of creation. Reduced to simple language, we have the explanation that an abundance of Life is bound to be the reward of those who would seek to expound the virtues of the one true God.

The "End of the World" Discourse:

To many, one of the more perplexing aspects of the Scriptures is the apocalyptic literature, which is commonly believed to consist of prophecy and to refer to worldly events of a physical nature. This is an unrealistic and unfortunate interpretation, for the real purpose of this form of writing is to symbolize events of a decidedly *spiritual* nature. Historical events are often interwoven with symbolism in an attempt to add authenticity to the text. In particular, the composer of an apocalyptic work will frequently endeavor to select an event in history which has already come to pass and then weave it into the story so that it will appear to be a genuine prophecy concerning the future— thereby obtaining much valuable prestige. (An example of this would be Daniel, which gives one the impression of being written in the sixth century BC and contains a glimpse into future events; while in reality it was composed near the middle of the second century BC, when such "prophecies" were already ancient history! Another instance would be what many religious leaders still consider to be a prediction of the destruction of the temple in Jerusalem in the year 70 AD— a "prophecy" written long after the event had occurred!)

Although often looked upon as a prophecy by Jesus involving the physical end of the world and the coming of "Christ," such passages really refer to *spiritual fulfillment* (i.e., the universal acceptance of the *Christ-idea*). A brief discussion of this symbolism follows:

"Truly, I say to you, there will not be left here one stone upon another, that will not be thrown down." (Matthew 24:2)

This supposed reference to the destruction of the temple does not concern a mere physical structure, but rather refers to the destruction of *false beliefs!* "Do you not know that you are God's temple and that God's spirit dwells in you?" (I Corinthians 3:16.) The "stones" mentioned here are used in their negative sense, signifying the destruction of false beliefs. (In addition, there is likely a secondary hope that this passage will come to be viewed as a prophecy foretelling destruction of the temple by the Romans— an event which occurred prior to the actual writing of this text!)

"Tell us, when will this be, and what will be the sign of your coming and of the close of the age?" (Matthew 24:3)

This passage really refers to world acceptance of the *Christ-idea*, and to the end or destruction of the materialistic mode of thought in each one of us.

"For many will come in my name, saying, 'I am the Christ, and they will lead many astray." (Matthew 24:5)

How true! The world has most certainly been plagued with the false teachings of so-called religious "authorities."

"And you will hear of wars and rumors of wars; see that you are not alarmed; for this must take place, but the end is not yet." (Matthew 24:6)

Along with several of the following verses, this passage means that universal acceptance of the *Christ-idea* will not be easy to a world so enveloped in materialism.

"And this gospel of the kingdom will be preached throughout the whole world as a testimony to all nations; and then will the end come." (Matthew 24:14)

Universal acceptance of the *Christ-idea* is quite impossible until all have received insight into the Truth.

"So when you see the desolating sacrilege spoken of by the prophet Daniel, standing in the holy place (let the reader understand)."

(Matthew 24:15)

This passage (see Daniel 12:11) refers to the taking away of the "daily sacrifice," which in turn was a symbol of the very principle behind the meaning of the seven days of creation.

"Then let those who are in Judea flee to the mountains." (Matthew 24:16)

In order to escape the consequences of failing to understand the significance of the seven days of creation, one must "flee to the mountains"—one must turn to exalted thought in order to rise above the situation.

"Pray that your flight may not be in the winter or on the sabbath." (Matthew 24:20)

Do not attempt to reach the Truth through the coldness and emptiness of materialism, nor through false theology.

"For as the lightning comes from the east and shines as far as the west, so will be the coming of the Son of man." (Matthew 24:27)

The light of the *Christ-idea* will one day be universal and it will appear to all. (This coming of the "Son of man" may be taken to symbolize the highest state of human perfection. Hence, the "Sons of men" must ultimately evolve to become the "Sons of God.")

"Immediately after the tribulation of those days the sun will be darkened, and the moon will not give its light, and the stars will fall from heaven, and the powers of the heavens will be shaken." (Matthew 24:29)

The Sun Moon and stars are used here in a negative symbolic sense and convey the message that *false systems or beliefs* are bound to fail, and must finally give way to knowledge that will indeed shake the "powers of the heavens."

"And he will send out his angels with a loud trumpet call, and they

will gather his elect from the four winds, from one end of heaven to the other." (Matthew 24:31)

Those who have sought the Truth (i.e., the "four") will be justly rewarded.

"So also, when you see all these things, you know that the end is near, at the very gates. Truly, I say to you, this generation will not pass away till all these things take place." (Matthew 24:33-34)

One possessing insight into the Truth will indeed know when salvation has come to pass; in fact, such virtue will be responsible for its coming to pass.

"As were the days of Noah, so will be the coming of the Son of man." (Matthew 24:37)

As the one is symbolic (i.e., Noah and his ark), so is the other!

"Watch, therefore, for you do not know on what day your Lord is coming." (Matthew 24:42)

The intended moral is that we do not know when death will overtake us, and that we should strive always after righteousness that we might not be caught unaware.

The Book of Revelation:

To those not versed in the system of biblical symbolism, the Book of Revelation must pose an enigma. As a matter of fact, so saturated with symbolism is it that it was very nearly never incorporated into church canon at all, since in the eyes of many early leaders— lacking the necessary key— it naturally appeared quite obscure. Regrettably, while most present-day theologians have been forced to admit that this work is largely allegorical in nature, they too have failed to offer a realistic interpretation.

Upon looking over the extensive symbolism so utilized, one finds that there is hardly a verse which does not contain one or more of the symbolic numerals! For this very reason it is extremely difficult to escape the conviction that its author has purposely overworked these "sacred" numbers in order to impress upon us the knowledge that the Scriptures were never intended to afford an account of mere historical events, but are instead concerned with the far more important realm of the spiritual.

In reality, throughout the entire course of this apocalyptic text there is an attempt to stress— *by means of symbolism*— the inevitable triumph of spiritual enlightenment over the darkness of mortal thought and mortal actions. This view is well illustrated in the very first chapter: "Behold, he is coming with the clouds, and every eye will see him, every one who pierced him; and all tribes of the earth will wail on account of him." (Revelation 1:7) The statement that "he is coming with the clouds" means that the *Christ-idea* has been expressed in the Scriptures through symbolism; and the words that "every eye will see him, every one who pierced him" conveys the message that the *Christ-idea* must one day be accepted universally, even by those whose thought is so darkened that they would "slay" the very idea. That "all tribes of the earth will wail on account of him" informs us that it will not be easy for the world to give up many cherished doctrines and a materialistic philosophy which would appear to afford immediate and tangible physical rewards for one's effort.

Inasmuch as this work is imbued with so many examples of symbolism, we must be content with but an introductory insight. Passing by such interesting instances as the diabolical beast in chapter 13, whose "number is *six hundred and sixty-six*" (symbolic of imperfect mortal thought); the "*hundred and forty-four thousand* who had been redeemed from the earth" in chapter 14 (symbolizing that those who sought to find and obey the One Infinite Cause will in truth be identified with God); and the ultimate downfall of that great city of Babylon (also in chapter 14 and referring now to the fall of Rome— *a symbol of mortal beliefs*— just as the falling of the walls of Jericho was used to symbolize the same annihilation of evil), we may take the following examples for brief comment:

The Messages to the Seven Churches:

The seven messages to the *seven* churches (Revelation, chapters 2 and 3) constitute an interesting specimen of biblical symbolism. Each message will be seen to correspond with the spiritual meaning imparted by one of the seven days of creation.

Message to Ephesus:

This church is accused of neglecting its "first love" (i.e., the first light of conscience, symbolized by the *first* day of creation.

Message to Smyrna:

We have here an illustration of the sense of the *second* day of creation, which connotes the separation and superiority of good over evil: "Behold, the devil is about to throw some of you into prison, that you may be tested, and for *ten* days you will have tribulation. Be faithful unto death, and I will give you the crown of life." (Revelation 2:10)

Message to Pergamos:

Criticism is directed in this particular instance against false beliefs, which are quite in contrast to the "dry land" state of thought of the *third* day of creation. Moreover, the "white stone" promised to those that overcome such false doctrines symbolizes the firm foundation offered by the "dry land" state.

Message to Thyatira:

The false teachings and principles attacked here may be compared with the *fourth* day of creation, stressing that God is a divine *System or Principle* which, eventually, must rule the world: "He who conquers and who keeps my works until the end, I will give him power over the nations, and he shall rule them with a rod of iron ... ," (Revelation 2:26-28). The "morning star" also promised is a further symbol of divine System or Principle.

Message to Sardis:

In this message we have criticism directed against many who are "dead" spiritually— quite the opposite of the *fifth* day of creation, which is characterized by the abundance and multiplication of Life: "I know your works; you have the name of being alive; and you are dead." (Revelation 3:1)

Message to Philadelphia:

Since this church had made an effort to utilize its limited potentialities, it is promised dominion over those whose thought is still darkened: "Behold, I will make them come and bow down before your feet." (Revelation 3:9) In addition, it promises intimate communion and identification with God: "He who conquers, I will make him a pillar in the temple of my God; never shall he go out of it, and I will write on him the name of my God." (Revelation 3:12) This is precisely the essence of the *sixth* day of creation— dominion and recognition of the immortal nature of both God and man!

Message to Laodicea:

This church is accused of thinking itself, without justification, to be in harmony with the *seventh* day of creation, which is characterized by a state of Perfection. "For you say, 'I am rich, I have prospered, and I need nothing; not knowing that you are wretched, pitiable, poor, blind, and naked." (Revelation 3:17)

The Seven Seals, Trumpets and Vials:

Since we cannot hope to go into the many instructive details of the *seven* seals, the *seven* trumpets and the *seven* vials, it must suffice to say that they are introduced into the text to symbolize a *sevenfold destruction of false beliefs*. For instance, let us take the sounding of the fourth trumpet as an example: "The fourth angel blew his trumpet, and a *third* of the sun was struck, and a *third* of the moon, and a *third* of their light was darkened; a *third* of the day was kept from shining, and likewise a *third* of the night." (Revelation 8:12). Understanding that it is false mortal beliefs that are being destroyed, and as a result recognizing the usage of the symbol "three" in a negative sense, it will be seen that there is an attempt to convey the message that eventually all false systems and principles are bound to fall. (The Sun, Moon and stars of the fourth day of creation are used as a symbol of System or Principle.)

The Red Dragon and the Woman:

The episode in the twelfth chapter of Revelation, concerning a pregnant woman persecuted by a "great red dragon," provides a typical example of the symbolism used to illustrate birth of the *Christ-idea* into the world. The woman wearing a "crown of *twelve* stars" is seen to be introduced as a symbol of the authority and universal extent of divine Principle; and the child which she was to bring forth is the *Christ-idea* being born into the hearts of mankind. Opposing such an event we have the "great red dragon," which is a symbol of a mental state of darkness and ignorance. (The *seven* heads and *seven* crowns together symbolize imperfection and the opposite of insight into the spiritual meaning of the seven days of creation; the *ten* horns are a counterfeit of the Ten Commandments; and the *third* part of the stars of heaven which its tail cast to earth is a symbol of a total lack of insight into Truth.)

However, as formidable as the "great red dragon" is depicted, and despite the efforts of imperfect mortal minds to prevent the birth of the *Christ-idea*, it is recorded that "the earth came to the help of the woman, and the earth opened its mouth and swallowed the river which the dragon had poured from its mouth." (Revelation 12:16). The intended moral, of course, embraces the prediction that eventually the world will turn against the "great red dragon" state of mind and come to accept the divine System of Principle we call God.

The "New Jerusalem":

The "new Jerusalem," or city "*foursquare,*" of the Book of Revelation (chapter 21), refers not to a physical structure, of course, but rather to the establishment of a new sense of government based upon Truth. The "foursquare" aspect is symbolic of much sought after Truth; while the *twelve* gates of the city and the *twelve* precious types of jewels which adorned the "foundations of the wall of the city" together symbolize that one may enter into Truth only through a demonstration of divine Principles. Upon gaining admission, we will indeed be blessed with a realistic perspective and possess something truly precious.

The fact that "I saw no temple in the city, for its temple is the Lord God the Almighty and the Lamb" (Revelation 21:22) conveys the important message that traditional organized religion and the idea of a "place of worship" must one day pass away, since the *Christ-idea*— as expressed through the medium of an individual's conscience— will be seen to afford the most successful route to Truth. Imparting a quite similar message is the statement: "And the city has no need of the sun or moon to shine upon it, for the glory of God is its light, and its lamp is the Lamb." (Revelation 21:23). For with universal achievement of a higher perspective there will no longer be any need to express matters in symbolic terms or to rely upon some external source of light, since the light of conscience will be both clear and adequate.

Words of Lasting Wisdom:

Regrettably, the New Testament continues to be very much misunderstood, suffering a great deal at the hands of skeptics and fanatics alike. But while it has been distorted and abused by so-called religious "authorities" masquerading under false pretenses, it nevertheless contains a wealth of constructive philosophy. The following are but a few of the many rich and enlightening passages taken from this unique collection of literature, and which all of us would do well to ponder:

"And you will know the truth, and the truth will make you free."

(John 8:32)

"For where your treasure is, there will your heart be also." (Matthew 6:21)

"But seek first his kingdom and his righteousness, and all these things shall be yours as well." (Matthew 6:33)

"Truly, I say to you, it will be hard for a rich man to enter the kingdom of heaven." (Matthew 19:23)

"For what does it profit a man, to gain the whole world and forfeit his life?" (Mark 8:36)

"And as you wish that men would do to you, do so to them." (Luke 6:11)

"Blessed are those who hunger and thirst for righteousness, for they shall be satisfied." (Matthew 5:6)

Reference:

[1[An informative account of the rise of the "Christ" myth may be found in a book entitled *Who Wrote the New Testament? (The Making of the Christian Myth)*, by Professor Burton L. Mack, ISBN 0-06-065517-8.

10

The Dead Sea Scrolls

Discovery of the Scrolls— Dating of the Scrolls— Contents of the Scrolls— Christian Origins— Suppression of the Scrolls

The issue of Christian origins is not a subject that our churches generally care to discuss. They prefer to let their flocks live a rather sheltered life, oblivious to important new evidence and to the impact of scholarly research over the past hundred years or so. In particular, they have endeavored to pretend that discovery of the Dead Sea Scrolls, during the middle of the 20th century, has posed little or no threat to traditional Christian beliefs. This could hardly be further from the truth.

Discovery of the Scrolls

Roughly 20 miles east of Jerusalem lies the northern tip of the Dead Sea, a land-locked body of exceedingly salty water extending southward almost 50 miles and with an average width of close to 10 miles. Constituting the lowest sizable body of water in the world, its surface is situated some 1,290 feet below sea level. Although less than 20 feet deep at the south end, its greatest depth is toward the northern edge where it drops to about 1,300 feet. Some six times saltier than the ocean, it is surrounded for the most part by rugged terrain and a series of steep cliffs containing numerous fissures and caves. It was from just such caves— near the northwest shore— that a wealth of ancient documents first surfaced in 1947, and findings continued over the course of the early 1950s.

At the extreme northwest corner of the Dead Sea, little over a mile from this unique body of water, is to be found the ruins of what was once a combination religious housing settlement and fortress now known as Khirbet Qumran. Dating back to the first half of the 2nd century BC, it was badly damaged and abandoned during the reign of Herod the Great, from 37 to 4 BC. Eventually rebuilt, it stood until finally destroyed for good by the Romans in 68 AD. Identified by historians as being occupied by a Jewish branch sect known as the Essenes, it was from just such a group that a vital key would be found to link the Dead Sea Scrolls to Christian origins.

The Essenes evidently populated a number of local communities, including nearby Jerusalem which was home to a large following. Over the years their once ascetic, reclusive and pacifist membership became infiltrated with a number of religious Zealots— men "zealous for the Law" of strict Judaism and antagonistic toward their Roman invaders, whom they regarded as an enemy to be expelled by whatever means were necessary. Although called Essenes by several ancient historians, inhabitants of the Qumran community often referred to themselves as being "followers of the Way," and ran their own social order upon the basis of a communistic economy— in which all wealth and possessions were shared and held in common. It was truly a distinctive philosophy which separated this group from the mainstream of Jewish society.

The actual story of the Dead Sea Scrolls began early in 1947, when a Bedouin shepherd boy noticed a small opening in the face of a cliff not far from the Qumran ruins. Crawling inside, he soon discovered a number of earthenware jars. It is believed that at least forty jars were found, some still intact and with sealed lids. Not realizing the true worth of the find, Bedouin tribesmen later removed a quantity of leather rolls (wrapped in decaying linen) from the jars. While it is not known for certain just how many scrolls were taken, a total of seven (more or less intact) documents eventually found their way into the hands of responsible authorities, along with fragments of almost two dozen others.

A brief description of these seven principal scrolls follows:

St. Mark's Isaiah Scroll is a well-preserved and complete copy of the Old Testament Book of Isaiah. Measuring some 24 feet in length and roughly 12 inches wide, it is the oldest intact copy of this ancient manuscript.

Hebrew University Isaiah Scroll, with most of the first half missing, is likewise a very old copy of this popular Old Testament document. Differing only a little in text from the rendition found in the Bible today, its leather is much deteriorated.

Habakkuk Commentary is a commentary relating to this minor Old Testament book, with new text applicable to events and ideas which were current at the time of the commentator. Measuring about 5 feet in length, it is just under 6 inches in width.

Community Rule, or "The Manual of Discipline," deals with rules of the Community and is also "sectarian" in nature (i.e., text that deals with contemporary ideas and issues). Roughly 6 feet long and about 10 inches wide (with its two separated leather sections joined), it also has a few pieces missing.

The War Scroll, or "The War of the Sons of Light with the Sons of Darkness," is a well-preserved apocalyptic and eschatological work which is similar in many respects to the Book of Revelation. This document measures about 9 feet by 6 inches.

Thanksgiving Psalms consists of a collection of psalms and parts of psalms not unlike those of the Old Testament. Considered as a form of literature, in the eyes of many scholars they are deemed to be slightly inferior.

Genesis Apocryphon, or "Aramaic Scroll," is written in the Aramaic language instead of in Hebrew. It contains chapters from the Book of Genesis, with expanded and embellished text.

Encouraged by the treasure of ancient documents found in what came to be known as Cave 1, it was inevitable that serious efforts would follow to explore every cave and crevasse in the vicinity of Qumran. In due course, a number of other caves, containing a large corpus of scrolls, were discovered. They were to set the stage for a scenario of intrigue and scholarly debate.

One particular find, known as the *Copper Scroll*, soon proved to be highly fascinating. Uncovered in March of 1952, in a site since designated as Cave 3, were two segments of a single scroll of rolled copper. With writing punched into the metal, it was obviously made to last and to withstand the ravages of time. Too brittle to be readily opened, it was not until 1956 that both halves had been sliced open and deciphered. It told of a vast treasure (mostly gold and silver) scattered over no less than 64 locations, with the focal point being the city of Jerusalem. Almost certainly, it had to refer to concealment of the Temple treasure in the face of an impending invasion from a large Roman army, which was dispatched in response to the Jewish rebellion of 66 AD. Unfortunately, the locations listed in the text have been rendered essentially meaningless with the passage of so much time, as reference points had long since faded into obscurity. To date, all attempts to find any portion of this huge inventory of wealth have ended in failure. Conceivably, the frequent usage of the "symbolic numerals" leads one to suspect that many quantities have been inflated— especially since the amounts appear to be far beyond what is likely to have been accumulated. (Examples of this symbolism may be cited as follows: "Beneath the steps that go to the east 40 cubits, a chest of silver; its totality of weight is 17 talents" "going down to the east 3 cubits up from the bottom, 40 talents of silver," etc.) If nothing more, it shows that the author of this scroll possessed insight into the unique system of numerical symbolism.

But the real treasure lay deep in the recesses of Cave 4, which was discovered in September of 1952, as it contained sizable portions of more than 500 separate documents. Incredibly, this cache was situated within a scant 50 feet of the Qumran ruins! Without a doubt, it was a regional depository for the community's extensive library of biblical manuscripts, hidden for safekeeping from an advancing Roman army. (It is conceivable, of course, that many of the cave documents may have come from Jerusalem.) Copies of virtually every Old Testament work, with the notable exception of the Book of Esther, were uncovered. (It would seem that the Esther "fairy tale" story, depicting a Jewess who is compelled to marry and become queen of a pagan king, did not find favor with the Qumran leadership who were highly nationalistic in outlook.) While numerous duplicate copies of much popular literature characterized this find, there were also many sectarian works and previously unknown manuscripts. It was in these latter scrolls— dealing with contemporary issues— that lay the real worth of the discovery, which soon led to a truly unprincipled campaign of deception and concealment. [1]

In order to deal effectively with such an enormous corpus of material, it was clear that the talents of many experts would be required to decipher and publish the 800 or so separate documents which had surfaced by the end of 1952. Accordingly, a team was assembled and actively engaged in this monumental task by the summer of 1953. Inexplicably, it excluded all Jewish scholars, being comprised chiefly of Roman Catholic priests under the leadership of Father Roland de Vaux— a highly opinionated and bigoted Dominican priest. Needless to say, from the team's very inception temptation to withhold or publish a biased translation— in favor of dogma inherent in traditional Christian beliefs— was bound to be both irresistible and inevitable!

Dating of the Scrolls

In the interest of "preserving the faith," it soon became clear to de Vaux that it would be dangerous to link the sectarian scrolls to the early Christian era. (Although none of the canonical Old Testament books posed a threat, there was a corpus of sectarian literature which dealt with contemporary issues and, as such, they could very well serve to undermine traditional beliefs.) In fact, the further one could distance them from the time of Jesus the better for orthodox theology. Thus it came to be that dating the Dead Sea Scrolls, or of associating them in any way with Christian origins, was crucial to the survival of church doctrine.

A number of diverse procedures are employed by scholars to date ancient manuscripts. In this instance, the following methods deserve discussion:

(1) Carbon-14 testing;
(2) Artifacts;
(3) Paleography;
(4) History;
(5) Internal evidence.

Radiocarbon dating techniques, first developed in 1947, offer a promising means of determining the age of an organic specimen. Based upon knowledge that the radioactive isotope carbon-14 (produced in the atmosphere by cosmic rays and incorporated into all living matter) has a half-life of about 5,730 years, before decaying into common carbon-12, it becomes feasible to derive an age by the simple expedient of assessing the respective ratios of the two isotopes. Improved considerably over the years, modern AMS (Accelerator Mass Spectroscopy) testing is far more accurate and wastes less material in the process. Carbon-14 dating is able to give reliable ages of organic material in excess of 50,000 years, although accuracy does fall off somewhat with age.

Shortly after their discovery, a piece of linen wrapping from the Scrolls was subjected to such a test and given a date of 33 AD— but with a plus or minus factor of some 200 years. Unfortunately, while this initial test promptly ruled out a medieval origin for them, it left too great a margin of uncertainty to settle the issue one way or the other. What really tends to boggle the mind is that, for more than 40 years, no additional and refined carbon-14 tests were ever performed upon these ancient documents!

Excavation of the Qumran ruins has left little doubt that it was once a dual purpose fortress and communal center with a library and, perhaps, a small scriptorium. Literally hundreds of coins were found, dating from early 1st century BC to the Jewish revolt of 132 to 135 AD under Simeon bar Kochba. An abrupt decline of coins from 68 AD coincides well with several historical records indicating destruction of the settlement by the Roman army at this time. The style and shape of the pottery jars, which contained the Scrolls, is also deemed by many scholars to be of 1st century AD antiquity.

While not an exact science, the paleographer can often date writing by the subtle manner in which alphabet characters and words change with the passage of time. Upon examining various samples, it was determined that there was indeed indication of a degree of evolutionary change consistent with a range of dates extending from 1st century BC to at least mid 1st century AD.

History, as a means of dating, may be invoked in order to provide a reason for hiding the Scrolls in the depths of barely accessible caves. In this regard, only three possibilities— all involving a Roman invasion— need be considered: Pompey in 63 BC; Vespasian in 66 AD; and Hadrian in 135 AD. (Actually, some documents pertaining to this latest era were found in caves more than 11 miles to the south of Qumran, including two letters signed by Zealot rebel leader Simeon bar Kochba. Relating solely to matters of the later rebellion, they are only of academic interest.)

The very contents of the sectarian documents themselves afford a measure of rather conclusive proof as to which of the three scenarios is the most probable. Among a wealth of internal evidence which may be adduced, through an examination of the published Dead Sea Scrolls, are the following examples:

Copper Scroll:

This scroll was almost certainly deposited for safekeeping (in Cave 3) in the wake of the 66 AD revolt. Since there was no Jerusalem Temple in 135 AD, from which a treasure might be obtained to hide, this late date is thereby ruled out. Moreover, there are few (if any) scholars who would seriously entertain the idea that this scroll could date back to the time of Pompey.

Roman Invaders:

Both the "Habakkuk Commentary" and the "War Scroll" documents allude to a Roman invasion— but one whose generals are commanded by a "king" or "monarch." Until the year 27 BC Rome was still a republic without an emperor. Thus the 63 BC date must be ruled out.

Herodian King Marriages:

In the entire corpus of Old Testament literature there is no mention of a Jewish king being forbidden to marry his niece. In two Qumran sectarian documents ("Temple Scroll" and "Damascus Document") this is strictly prohibited— revealing it to be a 1st century AD concept directed at Herod! Since John the Baptist was executed for vigorously condemning the marriage of Herod Antipas to his niece, this clearly rules out all but the 66 AD date.

Wicked Priest, Liar and Teacher of Righteousness:

These three unnamed characters appear together in a number of the Qumran sectarian writings, most notably in the "Habukkuk Commentary" and in the "Damascus Document." As we shall presently see, all three can now be identified with personages contemporary with Jesus and who lived for a number of decades after his death. Accordingly, such documents may be dated around the middle of the first century AD.

It will therefore be concluded that, beyond any reasonable doubt, all of the aforementioned sectarian documents are editions which were written within a limited first century AD time frame— most likely from the time of Jesus until the destruction of Qumran in 68 AD.

Contents of the Scrolls

Of the published sectarian Dead Sea Scrolls, which is the only type of literature capable of imparting any insight into Christian origins, the following documents are deserving of special comment and elucidation:

War Scroll:

Copies of this document were found in Caves 1 and 4, both of which are situated in the immediate vicinity of the Qumran ruins. This work was almost certainly written in response to an impending "Kittim" (Roman) invasion of their community. Depicting a symbolic war between the "Sons of Light" (themselves) and their adversary, the "Sons of Darkness" (the Romans), it expresses the notion of eventual triumph if their members are "zealous for the Law."

Instead of a somewhat monastic atmosphere characterizing Qumran— as orthodox theology had once supposed— it is now evident that this community had been infiltrated by large numbers of Zealots who were prepared to fight with great fervor. Indeed, since texts identical to several Qumran sectarian scrolls were found at Masada (a cliff fortress some 30 miles to the south, which was finally subdued by the Romans in 74 AD), it is clear that the two groups were essentially of the same sect.

Community Rule:

One of the original finds in Cave 1, it deals primarily with rules and moral code within the Qumran community which, it may be stressed, was not restricted to just the inhabitants of this one physical structure. On the contrary, there were probably many more "followers of the Way" (Acts 9:2) that lived in Jerusalem and other neighboring settlements. Glossing over assorted trivia, there are two rather remarkable parallels with New Testament writings which deserve mention.

The first and most striking observation involves what can only be termed a "communistic" society, in which it is stressed that all wealth and assets are to be shared equally by members of their community. A fundamental statute, within the text of this scroll, is the requirement of new members to "bring all their knowledge, powers and possessions into the Community." Another passage states that "his property shall be merged and he shall offer his counsel and judgement to the Community." This is precisely the gist of Acts 2:44-45, in which it is stated: "And all who believed were together and had all things in common; and they sold their possessions and goods and distributed them to all, as any had need." One can hardly escape the conviction that many of the original followers of Jesus (only later to be called Christians) were intimately associated with the Qumran community!

Yet another link between the New Testament and the "Community Rule" is the way in which they utilize similar phrases and expound similar concepts. Jesus' famous "Sermon on the Mount." for example, is common to both bodies of literature. (This fact will be seen to present somewhat of a dilemma for many Christian theologians who, in a desperate move to distance Qumran from Jesus by considering the Scrolls to be pre-Christian, would then have to concede that Jesus was merely echoing an earlier work!) Again, emphasizing their great "zeal for the Law," this document expressly states that anyone who "transgresses one word of the Law of Moses, on any point whatever, shall be expelled." This is essentially what Jesus is reported to have said: "Think not that I have come to abolish the law and the prophets; I have come not to abolish them but to fulfill them. For I say to you, till heaven and earth pass away, not a iota, not a dot, will pass from the law until all is accomplished." (Matthew 5:17-18). Once more, we are compelled to concede a connection between Jesus and Qumran.

Temple Scroll:

Believed to have been discovered in Cave 11 by Bedouins, this scroll eventually found its way into the hands of a local antiquities dealer. Six years later it was turned over to Israeli authorities and finally published in 1977. Appearing on the surface to deal with little more than trivia relating to furnishings, laws and rituals involving the Temple in Jerusalem, it does provide unexpected insight into certain facets of 1st century AD history and a link to another more valuable work.

The "Temple Scroll" reveals that the Qumran sect was not quite so divorced from the mainstream of Jewish religious life as once thought. To them the Temple was still their center of worship and its ruling theocracy was still very much their concern. In fact, it was the leaders of the "followers of the Way" who deemed themselves to be the true priests of the Temple, even if not designated as such by any Roman puppet king. Feeling justified by authority, it was merely a short step to write (or edit) scriptural text to reflect their own ostensibly superior views. Thus it came to be that text was composed forbidding a Jewish king to marry his niece. (Looked upon by the Qumran leadership as a foreign king, and not one "zealous for the Law," they championed text which would cast Herod Antipas in a bad light.) In any event, they performed the useful function of dating this scroll to within a relatively narrow time span.

As it turned out, this dislike of King Herod served to clarify a long-standing mystery regarding another ancient scroll known as the "Damascus Document." Discovered in 1896, in the loft of an old synagogue in Cairo, scholars had hitherto been unable to date it or to place it into context with other literature. However, with knowledge of the "Temple Scroll," they could see that it belonged to the Dead Sea library of documents. This connection was now firmly established by the very fact that, of the entire corpus of earlier biblical writings, only the "Damascus Document" and the newly-found "Temple Scroll" spoke of a prohibition against a king marrying his niece!

Damascus Document:

Portions of eight copies of the "Damascus Document" were found in Cave 4 at Qumran; while parts of two more copies were discovered in Caves 5 and 6, indicating that it was a popular work. Known to scholars for more than half a century, this document was given its name by reason of a reference to a place in the wilderness called "Damascus," where an unidentified "Teacher of Righteousness" is described as leader of a Jewish sect that remained "true to the Law."

What is so highly significant about the name of this place is that it is able to clear up a mystery with regard to Acts 9, of the New Testament, where Paul is reported to have been blinded for *three* days while on the road to "Damascus." Previously, it had been quite wrongly assumed that Damascus in Syria was the city in question. But this had always posed a problem, since an arrest warrant issued in Jerusalem would have been useless in such a foreign country! Thus there can no longer be any doubt that this communal residence (or fortress) of Qumran was originally called "Damascus" by the inhabitants. (As we shall presently see, this information has all the aspects of a "Rosetta stone"— one that is capable of revealing the truth about crucial events concerning the apostle Paul and Christian origins.)

Habakkuk Commentary:

This scroll was among the first that were found in Cave 1 at Qumran. In conjunction with a number of other documents describing the same trio of mysterious characters, the "Habakkuk Commentary" poses a very real threat to traditional Christian beliefs. There is repetitive mention of a conflict involving a "Teacher of Righteousness," a "Liar" and a "Wicked Priest." Who are they? What role do they play in the scheme of events just prior to the Jewish uprising of 66 AD? Due in no small measure to the research of Professor Robert Eisenman, it is now possible to finally answer such questions. [2]

Although initially there was a strong temptation for orthodox theology to picture Jesus as the "Teacher of Righteousness," this idea is beset by a serious flaw. In short, there is not the slightest hint— in all the Dead Sea Scrolls— that their leader was ever considered to be divine by his followers! If those in immediate contact with Jesus did not look upon him as the "Son of God," then how could Christians today be expected to uphold such a belief? As a result, this notion had to be abandoned and the mystery was left unresolved by Christian theologians.

The key to identifying the three aforementioned participants mentioned in the conflict is to be found in the "Liar," who is described as being a *traitor from within*— one who had been taken in by the Qumran community, accepted as an accredited member following an initiation period, and who subsequently defects. Recalling that Paul of Tarsus was greatly changed after his experience on the road to "Damascus" (Qumran), it is most logical to picture him as the "Liar." He does, in fact, meet all of the requirements, and he was in the right place at the right time. But, one might ask, in what sense did he defect? Quite simply, he was the first to preach that Jesus was anything more than a human being— that he was, indeed, an incarnation of God!

The identities of the other two, the "Wicked Priest" and the "Teacher of Righteousness," promptly fall into place. The high priest Ananas was the Temple priest at this time and is reported to have been an adversary of James— an early pre-Christian priestly leader— and therefore a "Wicked Priest." By association, it seems most likely that James is the unnamed "Teacher of Righteousness." (Conceivably, this esteemed title could refer to any current leader of the Qumran community.) In any event, James would appear to be the one so depicted in this instance.

Armed with the identities of the quarreling trio mentioned in such scrolls as the "Habakkuk Commentary" and the "Damascus Document," we are now in a position to proffer a revolutionary picture of Christian origins— one which is not likely to find favor with any facet of organized religion.

Christian Origins

As we have seen, canonical New Testament literature is not a very reliable guide when it comes to dealing with actual fact, since it is heavily biased in favor of imparting a specific religious message. Thus, in order to place the issue of Christian origins into true perspective, it is desirable to begin with what is known as historical fact, and then to extrapolate a probable scenario from the new evidence which is implicit in the Dead Sea Scrolls. To this end we shall commence with the lives of Jesus, James and Paul— the three who were involved initially in what later evolved into our present-day Christianity.

The Personage of Jesus:

Most likely born between 6 and 4 BC, probably not too far from Jerusalem, Jesus appears to have been well-versed in all aspects of biblical scripture. Highly antagonistic toward both the Sadducees and Pharisees, we are left with only the neighboring Qumran community as a possible source for his later education. Indeed, with the evidence strongly weighted in this direction, it can hardly be doubted that Jesus must have spent considerable time studying and conversing with members of this particular Jewish sect. Eventually, convinced that he had learned all that they could teach him, he then departed and began to preach on his own.

It is likely much more than a coincidence that at least some (if not most) of Jesus' original followers were recruited from among the Qumran membership. In particular, it is now little short of obvious that John the Baptist must have been intimately entwined with the Qumran sect— especially since he was so zealous in defending Qumran policies (i.e., the Herodian king/niece marriage affair) that he was executed as a result. By reason of his close association with John the Baptist, it must be concluded that Jesus was of the same school, so to speak. (Incidentally, the very fact that Jesus is not mentioned, in the entire corpus of Qumran literature, is a clear indication that while they must have known of him, this community did not hold him in high regard— most certainly, not as any kind of divine Savior!)

Unfortunately, an infusion of Zealot elements into the Qumran sect was to prove the undoing of Jesus. That he was indifferent or even sympathetic to their cause is attested to by the mention that "Simon who was called the Zealot" (Luke 6:15) *was actually one of Jesus' chosen apostles!* Also testifying to the rebellious company that he kept was the reported incident in the garden of Gethsemane, where Jesus' disciples are described as being an armed band. (Peter is said to have taken his sword and cut off the right ear of the servant of the high priest.) Another occasion in which violence was displayed is the episode of Jesus overturning the tables of the money-changers in the Temple (Matthew 21:12). Moreover, statements like "Do not think that I have come to bring peace on earth; I have not come to bring peace, but a sword" (Matthew 10:14) can only make one think of the militant Zealots. Yet a further indication of his involvement with this "zealous for the Law" sect is the fact that the Romans saw fit to crucify him— a punishment employed exclusively against political and rebellious offenders. Either Jesus was framed, or he was himself a member of this extremist group!

It is, therefore, to be concluded that there are two sides to the personage of Jesus. On the one hand, he would seem to have been an exceptionally wise and enlightened philosopher for his era. By way of contrast, he might well have become unwisely involved with a group displaying a little too much "zeal for the Law." (This association is clearly attested to in Acts 5:35-39, where Jesus' disciples are compared to earlier bands of militant Zealots.)

The Personage of James:

Mentioned in the Bible as a brother of Jesus (Gal. 1:19), the character of James has been quietly pushed into the background by church theologians. Indeed, much of the early literature relating to James is rather mysteriously excluded from the New Testament canon. Why? This is even more surprising when it is considered that James is described as head of the early church in Jerusalem— instead of Peter, who is supposed to be the "rock" upon whom Jesus would build his church (Matthew 16:18). The answer is one which completely undermines the very fabric of orthodox Christian beliefs: James was a staunch proponent and upholder of the "Law of Moses," *and as such would have nothing to do with the wild idea of considering Jesus to have been an incarnation of God!*

This interpretation explains the pronounced rift between Paul and James, since the former was actively preaching a story about Jesus' resurrection and divine nature. Unable to tolerate such an obvious fabrication, and as the "Teacher of Righteousness" head of the Qumran/Jerusalem sect, James is forced to object and to publicly admonish Paul. In turn, Paul's repeated assertions eventually gets him into serious trouble with James' zealous "followers of the Way," and he is only saved from death at their hands by Roman soldiers (Acts 21:32).

James' second adversary was the Temple high priest Ananas (depicted as the "Wicked Priest" in several documents of the Dead Sea Scrolls). Considered by James to be a puppet of Roman authorities, Ananas was reviled by Qumran "followers of the Way." Despised in turn by the "Wicked Priest," James was looked upon as an unwelcome rival of the Temple establishment. Moreover, by reason of suspected Zealot connections (real or imagined) James was likely deemed a threat to Temple security and a feared Roman reprisal. It was an atmosphere in which something bad was bound to happen.

When violence finally broke out the first major casualty would be James, who was severely beaten, eventually perishing (around 62 AD) at the hands of Ananas' hostile supporters. Incensed by this attack against their leader, the Zealot faction within the Qumran community subsequently managed to exact a measure of vengeance by assassinating the "Wicked Priest." Not long thereafter, all of Jerusalem exploded into violence, culminating in a general uprising and the Jewish revolt of 66 AD.

The Personage of Paul:

Born in the city of Tarsus (now in Turkey), Paul was of Jewish heritage with Roman citizenship. Well-educated, he was fluent in several languages. Never having met Jesus, he enters the biblical scene as an official entrusted with the task of arresting certain Qumran opponents of the Temple establishment. At this time little more than another radical Jewish sect, infiltrated with Zealots, there was no thought of this group ever becoming the nucleus of a new and dominating religion. Merely a simple expedition to "Damascus" (Qumran) in order to arrest "followers of the Way" ringleaders, this seemingly insignificant trip would serve to drastically change the course of history.

According to the text in Acts, chapter 9, Paul (called Saul, at this point in time) is in charge of a band of armed men en route to Qumran when a "light from heaven" supposedly knocks him to the ground. Whether it be pangs of conscience for previous misdeeds, or plain sunstroke, he is reportedly stricken blind for *three* days. When insight into the Truth (the symbolic meaning of the numeral "three") is imparted to him by a member of the Qumran community, he is completely changed. Allegedly spending *three* years at Qumran (Galatians 1:17-18), Paul is converted and accepted into their society as a legitimate member. He then departs to join James in Jerusalem, ostensibly most anxious to redeem himself by now preaching the sect's "zeal for the Law."

Evidently encountering a measure of distrust from some of James' associates (understandable in view of past experience), it is not long before Paul is farmed out to more remote locations— clear out of the country, in fact. This turns out to be a big mistake. Thrust into contact with other pagan religions boasting of a variety of most impressive deities— including semi-gods with magical powers— it is only a matter of time until Paul succumbs to a strong temptation to embellish his account of Jesus in order to be more competitive. In due course (perhaps after a little too much wine?), he finds himself making exaggerated claims and preaching a miraculous resurrection from the dead. The people just love his stories and he soon begins to make large numbers of converts. The ground is fertile and the promise of a heavenly life is exactly what the masses want to hear.

However, word of Paul's deification of Jesus eventually reaches James in Jerusalem. Paul is subsequently recalled and rebuked for his actions. A compromise of sorts is reached, to the effect that some aspects of the "Law of Moses" will be relaxed for non-Jews to join the church in the form of second-rate members. In return, Paul evidently agrees to tone down his preaching and stick to Qumran doctrine. He then embarks upon what must be described as a rather extended and incredibly successful mission— the only problem is that his many converts believe in a Jesus/Christ deity!

Again recalled to Jerusalem by James some years later (about 58 AD), Paul foolishly thinks he can still come to an equitable agreement with a church leadership that has by now little in common with Pauline philosophy. Confronted with the evidence, Paul is soon branded a traitor and a "Liar" by the Qumran community, in spite of the monetary contribution which he brings to the church from his recent converts. Ambushed by a mob of pious Qumran extremists, much less tolerant than James, Paul is barely saved from sure death by the intervention of Roman soldiers. Following a somewhat lengthy series of questioning, by various authorities, he is ostensibly shipped off to Rome to face some kind of trial, where he fades out of the picture.

But the die has been cast. In short order the post-Jesus church is superseded by a new and radical Christ/Jesus religion— one which very quickly spreads to surrounding nations, and which bears very little resemblance to the philosophy expounded by its alleged founder!

An Unplanned Religion:

It must now be abundantly clear, to all who would seek truth in an unbiased manner, that present-day Christianity is built upon a framework of deception and lies. Its major premises are seen to have absolutely no basis in fact and can only serve to disgust those who would exercise the power of human logic. The great irony of it all is that the theological structure of this rather peculiar Christ/Jesus/Savior religion was never intended by Jesus, who was simply trying to uphold the "Law of Moses."

Following in Paul's footsteps, it was not long before the new religion was further embellished by a succession of deceived and over-zealous theologians, scriptural writers and editors. Thus it was that Old Testament ideas of a "Messiah," who would lead the Jewish race to world supremacy, was replaced by a New Testament theology embracing the wildest of claims. First, there was the notion that a mere human being was God! Then, incredible as it might seem, it was claimed that through faith and a mystical blood redemption process man could be "saved" and pass to a "heaven"—completely bypassing the vital principle of free will and contrary to a firm cosmic requirement that destiny must lie in the hands of the individual!

To date, organized religions have relied upon blind faith and lack of knowledge in order to retain their flocks. Hopefully, with fresh insight into the Dead Sea Scrolls, it will not be long before nebulous faith gives way to true knowledge and reason.

Suppression of the Scrolls

Confronted by the wealth of material discovered in Cave 4, it was apparent that some sort of master plan would be needed to assess, translate and publish the huge find. Accordingly, the idea of forming an International Team of scholars and linguistic experts was soon conceived. With its headquarters at the Rockefeller Museum, located in the Jordanian sector of East Jerusalem (where Israelis were virtually excluded until the war of 1967), this team commenced work in 1953. Unfortunately, the membership of this team left much to be desired, and promptly set the stage for what would turn out to be subsequent deceit and scandal.

The head of this historic group, and editor-in-chief of all the published documents until his death in 1971, was Father Roland de Vaux. This was just about the worst possible choice that could have been made, as he was a highly zealous Roman Catholic priest who was extremely bigoted against Jew and non-Christian alike. He was also an appointed consultant (1955) to the Pontifical Biblical Commission, an ecclesiastical body whose sole purpose is to "preserve the faith"—an office which may be looked upon as our modern-day successor of the infamous Grand Inquisitor! Already, there existed both motive and opportunity to cloud the truth.

Appointed under de Vaux's authority, at this time, were the following original team members:

Professor Frank Cross:

From the Albright Institute in Jerusalem, and originally associated with the McCormick Theological Seminary in Chicago.

Monsignor Patrick Shehan:

Also from the United States and, at the time, the director of the Albright Institute and an esteemed Roman Catholic priest/scholar.

Father Jean Starcky:

An expert in Aramaic linguistics, and a Roman Catholic priest/ scholar from France.

Father Josef Milik:

A Polish Roman Catholic priest who lived in France, he was a close associate and confidant of de Vaux— so close, in fact, that he would later be the recipient of some of the most sensitive sectarian material.

Dr. Claus-Hunno Hunzinger:

A German expert who would shortly leave the team and be replaced by *Father Maurice Baillet*, another Roman Catholic priest from France.

John M. Allegro:

The only team member without specific Christian affiliation, a highly qualified British philologist with a number of publications in prestigious academic journals to his credit. (He would eventually be phased out of the team and replaced by *Professor John Strugnell*, a Protestant theologian later converted to the Roman Catholic faith and a staunch supporter of Christian dogma.)

Incredibly, no Jewish scholars were included as members of what was supposed to be an "International Team"—in spite of the fact that many were well-qualified for the task of translating the writings of their own ancestors. But such was the bigotry which prevailed at this crucial moment of Dead Sea Scroll history. This inexcusable bias was to continue for almost four decades, during which time the leadership of the team would pass from Father de Vaux to his chosen successor, *Father Pierre Benoit*, in 1971. Upon Benoit's death in 1987, it passed to John Strugnell, so that the theme of supporting Christian dogma would be preserved. Moreover, not only was leadership inherited, but even the rights and all unpublished translations were bequeathed to designated successors. In effect, a small clique of religious bigots had somehow managed to hijack for themselves a priceless legacy which rightly belonged to all of humanity!

From the very beginning, it was deemed essential to distance the Scrolls from Jesus and Christian origins. To this end, de Vaux and his companions in deception strove diligently to withhold evidence that certain of these documents were written shortly before the Roman/Jewish war of 66 AD— the war which necessitated concealment of the Scrolls in the dark recesses of the Qumran caves. Adopting a campaign of lame excuses and unprincipled delaying tactics, they were successful in suppressing full knowledge of their contents for many years. Fortunately, one member of the team was not willing to go along with their nefarious plan.

Of the International Team's membership, only John Allegro was really free to examine the Scrolls in an impartial manner. Needless to say, a difference of opinion was bound to arise; it was just a question of when matters would come to a head. By 1956, Allegro had managed to get into serious conflict with de Vaux and his minions. Given access to less and less material of any consequence, it became clear to him that there was intention to phase him out. However, before he died in 1988 he had succeeded in stirring up quite a storm. Publishing a number of informative and highly controversial books, he brought to public attention the International Team's many inordinate delays and exposed their suppression of all sensitive documents. It is reported that in early 1959 he wrote to Awni Dajani, Jordanian curator of the Rocke- feller Museum, expressing his fear that if any scrolls were to turn up which affected Roman Catholic dogma, the world would never see them. He was convinced that de Vaux was likely to send such objectionable material to the Vatican, where it would almost certainly be hidden or destroyed.

Regrettably, Allegro's courageous stand did not result in any immediate or substantial change of policy. Scrolls which should have been translated and made public in several years remained unpublished after several decades. Updated carbon-14 tests were yet to be made long after they were perfected to the point where they could have been most revealing. Photos of unpublished text were not available to impartial scholars until the early 1990s, and only then after a great public outcry and long overdue Israeli intervention, which finally resulted in the ouster of Strugnell as head of the team in late 1990.

But a rather interesting development of sorts is a story concerning Father Milik, initially a close associate and right-hand man of de Vaux. Given a "lion's share" of the most important sectarian documents, Milik returned to Paris with photos of this material— which he alone had been entrusted to decipher— prior to the 1967 Arab/Israeli war. He eventually left the priesthood and became somewhat of a recluse. One is tempted to wonder at this sudden change of attitude. What happened? Is it conceivable that, in studying this material, he found irrefutable evidence that completely undermined his Christian faith? If so, will the world ever learn the truth about such documents?

This very thought raises a disturbing possibility. In all the years of suppression, by a religious clique sworn to uphold and defend Christian dogma, just how many Dead Sea Scrolls were actually destroyed on grounds that they posed a serious threat to the faith? Alas, we may never know, as no independent observers were ever permitted to catalog these ancient manuscripts for well over a generation after their discovery. Posterity will always wonder how, in this supposedly modern age, such a shameful situation could have been allowed to develop.

References:

[1] An informative account of this deception is to be found in a book entitled *The Dead Sea Scrolls Deception*, by Michael Baigent and Richard Leigh, ISBN 0-671-73454-7.

[2] A scholarly work on James is entitled *James the Brother of Jesus*, by Professor Robert Eisenman, ISBN 0-670-86932-5.

Appendix 5

Evidence from the Tomb

Subsequent to revelations from the Dead Sea Scrolls, a rather startling discovery was made in 1980. However, it was not until some decades later that the full significance of this find became evident and made available to the general public in the form of a TV broadcast (mysteriously limited in exposure), and in a most informative book. (See: *The Jesus Family Tomb*, 2007, by Simcha Jacobovici and Charles Pellegrino, ISBN: 978-0-06-119202-9.)

Incredible as it may seem, when a tomb outside the old city walls of Jerusalem, in a hill region called Talpiot, was accidentally uncovered in 1980 by a bulldozer excavating a foundation for a housing development, the inscriptions on some ten bone ossuaries found within were given only a callous glance. In short order the contents were mixed and dumped into a common mass grave reserved for such excavated ancient human remains. (An ossuary is a small stone coffin in which were placed the bones of one after the deceased had been laid to rest in a crypt for about a year and the flesh had decayed to dust. Of particular importance, which serves to date the tomb, it is to be noted that this unique Jewish practice of a "second burial" was just a temporal passing phase over a span of a few generations— namely, from about the time of the birth of Jesus to the destruction of Jerusalem by the Romans and a major dispersal of its inhabitants in 70 A.D.)

Not only did one of these ossuary inscriptions read "Jesus, son of Joseph," but other ossuaries contained such biblical names as two "Marys," a "Joseph," a "Matthew," and one reading "Judah, son of Jesus!" Moreover, yet another was even inscribed "James, son of Joseph, brother of Jesus!" Who, in the light of such archeological evidence, could not conclude that this crypt was most certainly a Jesus family tomb?

One would surely have expected authorities to look upon this site as the greatest biblical find ever. Instead, it was allowed to pass with little fanfare and dismissed by many so-called "experts" as just a coincidence that the crypt contained so many names associated with Jesus. Clearly, none are so blind as those who will not see— for fear of totally undermining their cherished theological dogma!

As if to add an aura of mystery and intrigue to the discovery, one of the ten ossuaries found in the Talpiot crypt vanished before they could be transferred to a Jerusalem museum. Somehow, the ossuary bearing the inscription "James, son of Joseph, brother of Jesus" was literally stolen shortly after it was cataloged, and before it could be studied in depth. When it finally surfaced, some two decades later in the antiquities market, it was actually claimed to be a forgery by a number of supposed experts desperate to preserve the status quo. Only later was it proven legitimate, beyond doubt, through analysis of its patina. (A patina is the thin layer of dust and chemical residue that has accumulated over much time— a spectroscopic analysis of which could be almost as exact as a fingerprint.) In essence, the James patina was a perfect match for the other nine ossuaries found in the Talpiot tomb, and was indeed the missing one stolen years earlier.

What is especially troubling to Christian dogma is the inscription of "Judah, son of Jesus." According to traditional belief, Jesus is not supposed to be married, much less having fostered any children. A further anomaly, which was thought to argue against a Jesus family tomb, was the inclusion of a second "Mary," subsequently identified as Mary Magdalene, most beloved female companion of Jesus. But since only family members would be buried in a family tomb, it surely follows that this Mary must have been the wife of Jesus and the mother of his son, Judah. Needless to say, such revelations completely negate all ideas of Jesus being some sort of mystical god-like deity!

What are the odds of this tomb not being the actual burial site of Jesus and his immediate relatives? Estimates range from one chance in tens of thousands to one in several million, depending upon the criteria used and the method of statistical analysis. Most assuredly, there can be little doubt but what this find is by far the greatest archeological discovery of all time with respect to biblical history. But it is unlikely to be a subject which our churches will care to discuss, as it must lead to serious admission of error, loss of prestige, and be highly detrimental to income. Hopefully, in time, the masses will finally come to recognize the truth of the Jesus/Christ divinity myth.

Part Three:

Social and Economic Revelations

11

The Social Order

Society and Religion— Capitalism as a System— Communism as a System— The Democratic Principle

The nature of the social order under which we live is of utmost personal concern. It may mean existence in a state of freedom, security and happiness on the one hand; or on the other, it may subject us to poverty, injustice and virtual enslavement. But it is not enough to involve ourselves solely with the economic system which regulates our immediate lives. It is imperative to understand that a man is still a man and a woman is still a woman regardless of where or under what regime one lives, and that a transgression against any individual is of concern to all who are worthy of membership in the human race.

In the past, the penalty for failure to comprehend and overcome religious and economic difficulties has led to wars that could have been avoided through knowledge. A similar failure in this present age of atomic and chemical/bacteriological warfare could well lead to a catastrophic annihilation of humanity. We have now reached the point where modern means of destruction simply forbids us to condone future wars. Economic and social harmony must be achieved among nations before it becomes too late. It must, in fact, be perfectly clear that we can no longer afford to be indifferent to world affairs, and that there exists both a moral obligation and a dire necessity to acquire a realistic perspective with regard to such matters. A spirit of universal love and cooperation must prevail over petty nationalism and religious squabbling if there is to be any salvation for mankind.

Society and Religion

As an influence upon the minds and polices of humanity, the prevailing religious atmosphere has generally been of dubious value. Marginally beneficial at times, it has also proven to be a source of much confusion and has caused a great deal of needless suffering. While perhaps helping some souls to obtain a more civilized outlook upon life, history has most certainly revealed many undesirable side effects. It has frequently acted to hinder the advance of science and knowledge, thereby imposing a false and inferior philosophy upon the masses. It has consistently led to oppression of individuals choosing to obey conscience in their quest for Truth. Indeed, it has even served to instigate a multitude of wars! In spite of its considerable influence and claims to be supportive of the poor, it cannot be said to have done much in solving many of the world's problems which are clearly centered upon inequality of wealth.

Fundamentally, there exists a warped concept as to what really constitutes true religion. The lamentable truth is that, even today, there are large numbers of pious individuals who are deceived in their understanding of the requirements of *true religion*, and who are oblivious to the shortcomings of organized religion.

True Religion is in the Heart:

In reality, true and undistorted religion is simplicity in itself. It does not embrace ceremonies or rituals of any description; nor does it necessitate one's belonging to a church or religious organization.

Instead, *true religion is solely in the heart!* It consists simply of a desire to better our spiritual selves. All else is seen to be quite superfluous. Pure religion has nothing at all to do with accepting some traditional set of beliefs. Rather, there is a personal obligation to develop a genuine sense of universal love, obedience to conscience and a sincere desire to ever seek and live the Truth— regardless of the cost.

Paul, that dynamic biblical personality, in spite of his Jesus-Christ fixation, did express quite well the importance of nurturing a sense of *universal love*, in preference to other virtues like increased knowledge and a firm faith:

"If I speak in the tongues of men and of angels, but have not love, I am a noisy gong or a clanging cymbal. And if I have prophetic powers, and understand all mysteries and all knowledge, and have all faith, so as to remove mountains, but have not love, I am nothing ... Love never ends; as for tongues, they will cease; as for knowledge, it will pass away ...So faith, hope, love abide, these three, but the greatest of these is love."(I Corinthians 13:1-2,8,13)

A more realistic interpretation of the cosmos is certainly to be treasured; but then again it must be remembered that at death one's knowledge or particular faith "will pass away"—simply because such is not actually an integral feature (but at best only an attribute) of the real self or soul. Man's earthly and imperfect views must invariably die with him; but his sense of *love* can never be affected, and will subsequently determine his future manifestation.

It may come as somewhat of a shock to some devoutly religious souls to learn that the multitudes of ceremonies, rituals and dogmas presently held in esteem actually profit a man nothing; but it is nevertheless true, for they are the ideas and product of mortal man and clearly bear his mark. Indeed, all of the ceremonies and mass vocalization now fostered by our churches are essentially meaningless and to no avail. The only thing which really counts is the *sincerity* involved. It is, in fact, very often distracting and a complete waste of time to comply with the burdensome requirements of the various religious rituals currently in vogue.

If man would only realize that he owes no allegiance to anyone or anything except his own conscience, priestly influence would no longer bind him to outworn tradition grievous to be borne, and to insidious dogma enslaving him and depriving him of his rightful heritage: a degree of freedom which Truth alone is able to provide. (Likewise, nationalism, or a foolish and puffed-up pride of one's country, would reveal itself to be exactly what it is— a vain and rather stupid sense of self-centeredness which many unscrupulous leaders have been all too aware of in the past, and all too eager to foster for their own evil ends.)

Just as the world once found it difficult to accept the idea of one invisible God, and clung to materialistic idol deities, so there exists today a similar tendency to adopt a "physical" approach with respect to communion with God. For instance, such an outward show of emotion as mass praising and praying— a basic mainstay of orthodox religion— is an unnecessary and wasted display of reverence. All that is needed is that we should develop the constructive habit of retiring periodically from the mad rush and excitement of the world so that our inner thoughts may be free to express themselves and where, by serious contemplation, we may thus acquire a new and more enlightened outlook upon life. In brief, true prayer consists simply of *deep and serious thought!* Solitude and genuine meditation are indispensable to the success of true prayer, which is fundamentally a highly *personal* undertaking. This often misunderstood principle is well enunciated in words ascribed to Jesus:

"And when you pray, you must not be like the hypocrites; for they love to stand and pray in the synagogues and at the street corners, that they may be seen by men. Truly, I say to you, they have their reward. But when you pray, go into your room and shut the door and pray to your Father who is in secret; and your Father who sees in

secret will reward you." (Matthew 6:5-6)

Another common misconception regarding prayer is the belief that one's prayers could influence God into granting favors— including intervention on behalf of others! Surprisingly, few appear to have stopped to weigh the momentous consequences of accepting such a view. Can we not see that no intervention of any sort is possible—without contradicting the very laws of nature? Moreover, is it not clear that the entire concept of God intervening to assist one, at the wishes or prayers of another, is nothing short of blasphemy against the just nature of the spiritual cosmos, which demands that one be held responsible for one's own actions? Can we not see the injustice of the very notion, and understand that if the desires of one could so induce God to find favor with another it would lead to absolute chaos? Those instances in the Bible where it is advised to "pray for those who persecute you" were really intended to convey the following message: *Seek always to improve your level of righteousness by thinking unselfish thoughts.* (The noble act of thinking of the welfare of another is itself an indication of an increased level of righteousness, and is thus an action which is to be encouraged.)

The conclusion, therefore, to be drawn from the idea of praying for others is that while it can obviously have no effect whatsoever in securing God's intervention, it can nevertheless be of definite value *to the one doing the praying!* Prayer is indeed seen to be a personal undertaking.

Society and Organized Religion:

Although criticism is something to which we all have a natural aversion, the fact remains that it is often essential if we would seek to advance the human race. In the past, organized religions have been guilty of propagating so many erroneous doctrines that one might well suspect they may have done more harm than good. Nor can the influence of many churches be considered an asset in this present day and age, since a host of false dogmas may be cited which pose serious obstacles to progress.

Some current examples of the public mischief wrought in the name of religion include a senseless objection to the practice of birth control, along with a totally unfounded attitude which has arisen against any act of abortion. Not only is it abundantly clear that the problem of a population explosion must soon necessitate universal acceptance of birth control as the sole alternative to utter chaos, but it is amazing that such an absurd doctrine could persist in this supposedly modern age. Invariably, the ludicrous dogma of papal infallibility has precluded all hope that any pope will be able to rectify past acts of stupidity without pulling down the whole edifice of Catholicism!

Equally surprising is why anyone would tolerate the idea that a microscopic fetus is as valuable as a human life. Obviously, this theological blunder must be traced to a misunderstanding of how our souls are incarnated. For instead of becoming manifest at the moment of conception, it must now be clear that this spirit does not enter the body until some crucial— and much later stage— in the brain's development. Thus a tiny immature fetus has no greater value than any other part of our physical structure, and the traditional objection to abortion (at least early stage) is seen to be quite baseless. (Actually, the question as to when the soul enters the body is really of little more than academic interest, since the aborted spirit of a minute fetus has not yet acquired the faculty of memory and experienced time— hence for all essential purposes, *it has never been reincarnated!* The fact that it must promptly enter into another embryo, of like status, must serve to preserve Cosmic Justice.) It is surely far better to be born healthy to parents who truly desire a child, and who are in a position to care for it, than to risk being born either unwanted or deformed!

For centuries a major criticism of orthodox religion centers on its negative attitude regarding the subject of evolution. The open hostility of the Roman Catholic Church toward the revelations of the now renowned Galileo, for instance, is a well-known fact to history students and affords a typical illustration of the blinding influence posed by church dogma. Today, science is no longer the frail child of early times; it has grown up and become a powerful factor in our lives. But even so, and despite the fact that it has revealed dogma after dogma to be nothing but a vain and foolish product of man's imagination, in general our churches have yet to overcome their reluctance to accept freely the findings of science.

The initial reaction of theology to the theory of evolution was one of bitter opposition; and even today this medieval prejudice has by no means disappeared. On the whole, the present attitude of the churches toward evolution ranges from one of unhappy (and no more than partial) recognition, to an outright rejection and a last desperate attempt to cling to an outworn and unethical doctrine of Redemption. The fact that the basic message of the Scriptures is in full agreement with the principle of evolution— which is really a spiritual process fundamentally— appears to have escaped notice. Let us hope that our theologians will soon come to understand that evolution is not to be feared. On the contrary, when properly understood it will prove to be their greatest ally.

Regrettably, any problems over evolution do not end with a mere difference of opinion on how life and the world came into being. Actually, the trouble is just beginning. Our churches fail to consider that their false teachings are in direct conflict with the minds of a younger generation exposed to the now-powerful influence of science. Unrealistic and obscure doctrines can only be expected to induce widespread disillusionment among those who have been compelled to accept evolution as a truth by reason of force of evidence. Furthermore, since theology has often branded evolution as an atheist philosophy which denies all hope of a future life, it must follow that many of those accepting evolution *will really have been persuaded by these same churches to adopt a materialistic outlook!* It is to be trusted that theologians will soon come to realize this, and that the dynamic principle of evolution will receive due recognition.

The duties of any worthwhile religious body are unmistakable with regard to the State. It must not seek to use its influence to impose, in any way, its own particular doctrine upon the life of a nation to the exclusion of rival views— *complete freedom of religious thought must be maintained at all costs!* But neither can it afford, without betraying its moral responsibilities, to refrain from criticizing any system of government which does not comply with the principle of true democracy, and which fails to provide an environment conducive to the development of the human soul. Nor must it neglect to press for the security and freedom of all, a higher degree of justice and a fair distribution of material wealth. The social order exerts far too much influence over the life of an individual to be ignored. It is often just as essential to strive to replace a state of economic rivalry and distrust with one of mutual love and harmony, as it is to preach a more desirable personal moral philosophy.

There is, of course, one basic explanation for the many failings of organized religion— *namely, their dogmatic structure!* In short, their strict adherence to dogma prevents them from accepting a new and superior point of view— if it should happen to conflict in any way with tradition. This is the one fundamental weakness of our churches, and therein is to be found the reason why they must one day submit to the advance of knowledge. For it is quite obvious that they cannot continue to support present dogmas much longer in the face of rapidly mounting criticism; and once they are forced to launch a serious and unbiased investigation into the mysteries of the universe it must follow that organized religion as we know it will cease to exist. Religion will now have become a personal responsibility— which it has always been.

Thus the value of an open-minded conscience is seen to be of truly fundamental importance. One may graduate from all of our planet's recognized centers of higher learning, be exceptionally versed in tradition, hold a lofty position and respected highly of men— yet still possess a perspective in life inferior to that of one who can boast of no such qualifications, but who is nevertheless blessed with much more: *the priceless advantage of free thought and the fortitude to follow conscience!*

Capitalism as a System

There are two fundamentally opposed economic ideologies upon which a social order may be based— specifically, capitalism and communism. In turn, the principle of *democracy* is of overwhelming importance to the proper functioning of either system.

While certain aspects of capitalism go back in history to the first ideas of private ownership, it was only with the Industrial Revolution of comparatively recent times that this system really began to assume its present structure. Evolving, upon occasion, by the slow process of trial and error, many capitalist principles are merely a passing phase in man's quest for economic harmony. As a result, there may be witnessed considerable variation in the extent to which certain features have been retained or modified by different nations. In many instances ideas have been incorporated which might best be described as attributes of another economic system. An example of this would be attempts to inject at least some measure of stability and security through government- sponsored pension plans and insurance against the specter of unemployment. A few of the more progressive capitalist societies have also adopted schemes of health insurance. Thus, it will be seen that the harshness of a purely capitalist economy has, to some degree, been improved by social reforms in which the interests of the public are not completely overlooked. In effect, one nation may be observed to embrace a number of socialist principles although still retaining an economy which is essentially capitalistic in nature; while another government may exhibit little change from the primitive notion of equating authority with personal wealth.

In spite of these modifications, however, there remains many important features of this economic system that deserve criticism and which cannot be remedied until such sweeping reforms have been made as to literally convert it into a radically different social order. For reasons of ethics alone capitalism must stand condemned, as it is seen to be characterized by an unmistakable spirit of materialism and just plain selfishness. On economic grounds it must also be rejected, if for no other reason than experience has proven that it will not work in a fair and equitable manner. Moreover, left to itself, it will invariably lead to the spectacle of imperialism! (A major cause of wars since the rise of capitalism, imperialism may be defined as "a need by the wealthy to invest surplus funds in business enterprises abroad." Expecting to reap much greater profits from such investments, there is naturally a strong temptation— and incentive— to control the governments of the various nations involved. Historically, the First World War may be cited as a classic example of a war that was foolishly inspired by imperialistic capitalism.)

The motivating principle of capitalism is material wealth, the acquisition of which is held to be of prime importance. All actions, policies and production of goods by industry are prompted by the idea of *personal profit*— quite irrespective of whether they are justified ethically! Instead of a spirit of cooperation there will exist a situation in which one is forced to compete against his neighbor. As a result of the lack of planning which prevails, men are frequently compelled to seize whatever means of employment the system chances to provide— whether such employment be an asset to the community, a senseless waste of time and resources, or even a liability to society itself. Furthermore, it strongly encourages an unfair distribution of wealth, with all the resulting injustices and evils which invariably arise when certain individuals are permitted to acquire wealth and power *at the direct expense of others who must do without!*

Nor can much be said in favor of the traditional law of supply and demand as a criterion for establishing prices, which should really be focused upon the cost of production. The idea of trying to extract as much as possible from another is clearly rooted in selfishness. In addition to being morally wrong, it is also far too conducive to unfair fluctuations and extortion to serve as a basis for the ideal economic system.

The very nature of capitalism is such that it is plagued with periodic recessions and depressions, the only alternative to which has been huge peacetime military expenditures— or the specter of war, an even more deplorable state of affairs which the system also tends to foster. This is an established fact borne out by numerous instances in both past and present events, and it is of little credit to our generation that so many should condone this economic weakness, with all its serious faults, through lack of knowledge. Business cycles in which prices, wages and employment fluctuate in a most undesirable manner are not due to chance, but have very definite causes arising from defects inherent in the system. For instance, is it not a strange coincidence that the Great Depression of 1929 to 1939, which was only terminated by a war, was worldwide in scope and embraced virtually all capitalist nations?

But why are capitalist societies flung periodically into a state of poverty amidst plenty? Why must people be thrown out of work, lose their homes, their health and their very perspective in life? Quite regardless of excuses which may be offered, there is just one basic underlying cause, and this may be traced to *inability of the public to purchase all the goods produced by modern industry.* Poverty is seen to exist, not due to a lack of goods, but rather to the fact that *there is insufficient purchasing power in the hands of the masses!* It is a fundamental rule of economics that the supply of money in free circulation must be consistent with the value of available consumer's goods (i.e., homes, furniture, food, clothing, automobiles, etc.). Failure to place adequate purchasing power in the hands of the common people must soon lead to a state of depression. Conversely, a state in which there is more money than goods invites inflation. In order to achieve a stable economy this balance must be maintained at all times, and inability to do so is a clear reflection of unsound features of the prevailing system.

Under normal circumstances it is merely a question of time until a capitalist economy will find itself in such difficulties that, unless additional purchasing power is soon injected into the system, it must surely plunge into the depths of a depression. This chronic inability of the public to purchase all of the available consumer's goods may be traced to *an unfair distribution of income* that sees large sums of undeserved money ending up in the pockets of a privileged class owning the means of production. Should this wealth fail to redeem a corresponding value of consumer's goods, the system will then be in danger of collapse.

While it is possible to restore some of this lost capability by way of plowing back excess profits into capital investments (i.e., new factories and additional means of production), it is also a fact that there is no way much of this money will filter down to the needy majority. There is a definite limit as to how many new manufacturing plants may be put to proper use— especially when the public is unable to purchase all the goods produced by their existing factories. (Under such conditions the threat of imperialism begins to loom on the horizon.) What is required in times of high unemployment is not tax cuts to business corporations and the rich (ostensibly to encourage capital investment), but an *increased money supply to the poor masses so that they might purchase a backlog of unsold goods!*

The very nature of a business slump is such that, once started, it tends to gain momentum. Once production is restricted due to lack of sales, large numbers of employees will be thrown out of work, thereby serving to add to the trouble by further reducing the ability of consumers to purchase. With high unemployment, wages and prices begin falling and hoarding of what purchasing power there is will appear, since during a recession businessmen and consumers alike will hesitate to buy today if prices will be lower tomorrow. Banks become panicky and commence calling in loans, forcing many borrowers into bankruptcy. As a direct consequence, debt-burdened merchants are compelled to unload their stocks of unsold goods on the market at sacrifice prices, driving both prices and wages down still further. In the face of widespread unemployment, millions will be unable to meet payments on mortgage and other loans, which still bear the high interest rates from periods of earlier prosperity. Needless poverty and misery abound in a land which is capable of providing a rich and enjoyable life for all, simply because the very structure of the economic system is at fault! (It used to be that capitalist business cycles were characterized by periods of either recession or inflation. As of late, our governments seem to have found a way to give us both evils at the same time!)

What forces are responsible for eventually lifting a nation out of the depths of a depression? Alas, history has shown that the solution most frequently adopted is often worse than the original complaint. A state of economic prosperity usually begins, of all times, *immediately after a nation becomes involved militarily!* The reason for a war (or threat of war) working such a miraculous cure is not very difficult to find. Not only does it create an acute demand for expensive weapons of destruction, but it also absorbs multitudes of hitherto unemployed persons into the armed services. Large sums of money soon finds its way into the economy of a nation because the wages paid to produce munitions— which constitute a complete waste of labor to the consumer— *represent a pure gain in the amount of cash available to purchase consumer's goods* The trouble is not now a lack of money, but often too much of it (although most savings are still in the hands of the privileged few) and unless appropriate taxation is imposed, runaway inflation is bound to follow.

Thus, a sure cure for a depression is to resort to war. If this is not feasible, then a major armament drive, in which large numbers of young people are seduced (or forced) into armies and the public is made to bear the burden of supplying expensive modern weaponry, will prove just as successful in saving the system from economic collapse. While clearly disgusting, such expedients will nevertheless serve to reduce unemployment and inject badly needed purchasing power. But what sort of an economic system is this which works best during a time of war, or when a fortuitous excuse can be found to rearm? Clearly, unmodified capitalism leaves much to be desired.

Communism as a System

In spite of all the negative publicity given to communism over the years, there are relatively few inhabitants of capitalist nations who can be said to possess a realistic insight into many aspects of this system. But such ignorance is perhaps to be expected in countries where control of information is largely in the hands of factions quite strongly opposed to the principle of public ownership. As a matter of fact, it is questionable whether present condemnation and hatred of communism has been fostered as much by those genuine in their sympathy for injustices under so-called communist regimes, as it has by others who have much to lose should it ever be decided that industry and wealth rightfully belong to all, and not to a privileged few.

But admitting that reports about communist-styled nations are often "edited" in a highly biased manner by capitalist news media, the fact still remains that as a social order discontent has generally been the rule and not the exception. The idea in itself of all individuals sharing the resources of a nation constitutes a very noble philosophy, and one that can hardly be criticized on grounds of either ethics or economics. However, valid criticism arises, not as a result of public ownership, but rather because there is frequently a distinct lack of such! While it may be true that industry and commerce have been taken out of private hands and placed in control of the State, it is also a fact that in many instances the State has not been the people, but a clique of authorities terminating in a virtual dictatorship. For this very reason, it is not even certain that a true state of communism has ever existed, as this principle actually implies public ownership and control by the people themselves. (A more accurate description might, therefore, be seen to embrace such terms as "quasi-communism" or "totalitarian communism.")

It is here, in the inability of citizens to have a voice in formulating their own policies, where the real objection to so-called communist nations lies. Whenever a body of people permit them-selves to be dominated by a clique who will not allow criticism of their actions, it is more or less inevitable that concern for personal feelings and even human lives will be placed in jeopardy. (A very good example of this would be Stalin, a former Soviet dictator who was almost as nefarious as Germany's Hitler.) Instead of regarding man as a sacred spiritual entity placed here on this planet for the express purpose of evolving spiritually, such a totalitarian regime will tend to look upon one as important and an asset only to the extent that his views coincide with the aspirations of the State— the will of a virtual dictatorship—and insofar as his labor may be utilized by said State. A person will be brought up to accept authority and compelled to accept values as emphasized by "leaders" who have little sense of values. *Respect for an individual's rights and the sanctity of conscience is in danger of falling to a new low!* It remains for the future to decide whether present totalitarian communist leadership will overcome existing objections and allow freedoms so sadly lacking in the past.

And yet, as oppressive and dictatorial as may have been certain quasi-communist regimes, one slight but vitally important change would make all the difference in the world. If only it was recognized that the State exists for the people (and not the people for the State) most of the many inadequacies now so evident in so-called communist societies would soon vanish. In short, establishment of a genuine democracy is very likely a dire necessity before objections to current alleged communist governments can be overcome.

Unfortunately, reaction of the major capitalist nations toward the plight of such oppressed masses has been disappointing. Not only have they frequently exhibited an attitude of indifference, but there have been many instances in which they have actually given military support to reactionary regimes opposed to social reform! The only sure way to halt the expansion of totalitarian communism is to assist undeveloped countries in their quest to more fairly distribute wealth. Failure to do so is bound to produce serious consequences, and any infamous role played by our capitalist governments will be duly acknowledged by posterity!

The Democratic Principle

While the principle of democracy must be regarded as crucial to the functioning of a just society, *it cannot be said that a genuine democracy is likely (or even possible) under a strict capitalist economy!* It may be true that the privilege to vote and to elect representatives constitutes a definite measure of democracy, but this is hardly a guarantee of an acceptable social order, nor evidence of the existence of a real democracy. On the contrary, while this principle is still in its infancy, the results so far can only be described as less than encouraging. Essentially, success of the democratic process hinges upon a combination of two factors:

(1) The *desire* of a nation's inhabitants to work for a better social order.

(2) The extent to which the public is given *unbiased* access to information.

The privilege of casting one's vote is a personal right to be cherished, posing a moral responsibility to the individual. But since the vote of an unintelligent and indifferent citizen carries just as much weight as that of one who is far more enlightened, it is wise to refrain from voting blindly. (It is surely more prudent to limit the right to vote to those capable of passing a stiff test designed to weed out the incompetent. Ability to discern *principles* is of the utmost importance in an election, and there is little merit in basing intelligence solely upon some arbitrary age. The privilege of voting is a right that should be *earned!*) To vote for a candidate simply because he or she has received much favorable publicity is surely a grave mistake that could very easily rebound against the voter. It is undoubtedly a great shame and an injustice that the more spiritually evolved must suffer for the incompetence of others, but such is life in our present mundane phase of evolution.

The reason why capitalism is so unfavorable to the existence of a sound democracy is due, in large measure, to the serious inequality of wealth which it fosters. Inasmuch as an individual's perspective owes much to the influence of one's environment, it is to be expected that such media as newspapers, magazines, radio and television will constitute a most important source of persuasion. But whose views do these influential organs reflect? Almost without exception they express the *personal views and aspirations of their owners*, who are invariably limited to wealthy individuals who manage to see eye-to-eye with big business interests—for the simple reason that they are big business! If it should suit their purpose they are able to tactfully withhold or even color news items that would tend to cast an unfavorable light upon their ambitions. On the other hand, they are able to magnify, often out of all proportion, certain trivial occurrences that happen to coincide with their policies. In short, by controlling the main avenues of information it is frequently possible to restrain those passions it desires to suppress, and to incite those passions that are likely to prove advantageous. (Ironically, before the advent of mass media, and the necessity of purchasing much flattering publicity, the democratic process stood a better chance of working.)

As regrettable as this one-sided control of information may be, the trouble extends to the actual choice of political representatives. Here money also plays a predominant role. It so happens that one lacking financially has little hope of success without receiving publicity and monetary assistance from some exterior source. As a result, it is not in the least surprising that most of our politicians and political parties owe allegiance to big business interests, whose selfish motives have led them to provide such contributions. But *nobody is likely to give something for nothing!* It is, in fact, no accident that genuine democracy is little more than an illusion and that, on the whole, our governments have failed to properly reflect the wishes of its citizens. It may well be considered a disaster of our times that so many freedom-loving nations hold periodic elections in which voters have little or no opportunity to elect a truly acceptable government—simply because there is not a sufficient number of worthy candidates from which to choose!

This point is amply demonstrated upon examination of the supposed "rivalry" between certain political groups. With few exceptions, elections in most major capitalist nations feature two of person who is most likely to possess the undesirable trait of selfishness!

Nevertheless, in spite of the many obvious economic, social and political defects of capitalism, the public is not without a definite measure of responsibility. As long as a nation has the freedom to vote, regardless of any unfair influence imposed by wealth, it is within its power to overcome such an obstacle and to create a society of its own choosing. Social reform can be obtained, but only at the cost of much effort and through the acquisition of an enriched perspective on the part of the masses. If we truly desire a better social order we must not be afraid to work and sacrifice to achieve this noble end.

Note:

One viable solution to the voting problem, confronting many supposedly democratic nations, is to drastically limit the amount of money spent on behalf of a political candidate. There is simply no need (or justification) for repetitious TV and newspaper ads costing enormous sums. Nor does it make any sense to display a multitude of signs proclaiming a candidate's name. (The mentality involved in the premise that the more signs the better is so illogical that it questions a voter's intelligence to discern values!) All that is necessary to express the views and policies of a political hopeful is an inexpensive brochure of no more than a dozen pages or so, which could be mailed or delivered by hand to voters. Strict enforcement of such a limitation would certainly serve to "level the playing field" and allow a worthy candidate (with little money) a much better chance to be elected.

12

A "World Society"

A Charter of Human Rights—Social Reform—Economic Reform--Global Unification

The sanctity of the human soul has all too frequently been ignored by governments guilty of denying essential rights under the pretense of conserving freedom. What is so desperately needed is a *World Constitution* effectively guaranteeing certain fundamental human rights— namely, a set of rules respecting the dignity of man and backed by the might of a well-armed and dedicated military presence operating within the structure of a revitalized and reformed New United Nations.

These rights deserve a measure of discussion, along with a number of suggestions for achieving global harmony.

A Charter of Human Rights

The primary purpose of our existence is to evolve spiritually, through the process of overcoming an inherently selfish nature. *Any interference with this sacred obligation must be considered to be a violation of basic human rights!* Reduced to simple terms, the State must be charged with the preservation of four cardinal freedoms. These freedoms may be defined as follows:

Freedom to follow conscience:

There are many ways in which governments have acted to the detriment of a soul following conscience in genuine pursuit of Truth. By far the most serious invasion of privacy lies in the despicable practice of forcing an individual into military service. In the guise of upholding freedom, one is required to relinquish freedom of both mind and body. Moreover, at a command, he is expected to openly defy conscience by committing a transgression against his fellow man. Without a doubt, this can only be viewed as a flagrant example of enslavement! Submission, in such a case, is tantamount to providing both ammunition and temptation for any unscrupulous leader to wage warfare. It is a fact that if all who are drafted for military service refused to enlist, no major war would be possible! If nations were really sincere in their professed desire for peace, they would see the wisdom of outlawing compulsory military servitude. Under the auspices of a much reformed United Nations, a worldwide ban on this outrageous intrusion into one's life could very effectively prevent future large-scale wars.

Linked closely to this crucial respect for conscience is freedom of speech and freedom from religious dogma. *In no way is the State justified in interfering with the free expression ideas or knowledge!* Indeed, if the prevailing social order is incapable of withstanding criticism, it is a sure sign that such a government is in dire need of revision.

Freedom from want:

It is to be regretted that an unprincipled government can have a most adverse influence upon the physical body housing the soul of man. With selfishness at the root of all evil, it is inevitable that spiritually immature leaders should seek to deprive others of a fair share of their rightful heritage. It is a fundamental law of arithmetic that whenever someone receives more than an average portion, so another must go short. Since there is only so much wealth available for distribution, it may be deduced that in the making of every rich millionaire many will be thrust toward poverty. Granted that some inequality of material wealth is inevitable (or even desirable), but present extremes of wealth are quite unacceptable and constitute a prime cause of much of the world's unrest.

The obligations of a truly progressive State are quite clear with regard to the physical well-being of its citizens. While assuring a just share of material goods in return for honest labor, it is also duty-bound to provide opportunities for gainful employment. Furthermore, in addition to acknowledging responsibility for the support of the aged and infirm, it must display genuine concern for the health and welfare of all. In short, no individual should be allowed to suffer needless hardship for lack of sufficient funds.

Freedom from discrimination:

Discrimination may take many forms, with underlying causes ranging from outright greed to simple ignorance. No respectable social order can afford to condone injustice, whether it be related to crime, wealth, religion, politics, education, color, race or sex. In an atmosphere conducive to spiritual progress, every attempt will be made to eradicate inequality wherever it is seen to exist. The very foundation of a truly *Cosmic Perspective* is based upon developing an attitude of unselfishness— essentially, a feeling of universal love that would consider all humanity as equal.

Yet another aspect of discrimination is one whereby governments think they have a divine right to admit or exclude inhabitants of our planet from crossing some nationalistic boundary. It is little short of imprisonment to prevent an individual from leaving a country, and a definite violation of human rights. On the other hand, while there may be times when there is justification for a nation to impose some limitation upon permanent immigration, such a restriction should not apply to those whose presence would not impose a financial burden. No portion of our globe belongs to any one group or race, and the goal of an evolving world society is not to restrict, but to facilitate exchange of culture.

Freedom from aggression:

Every individual has an inalienable right to be protected from all manner of undue persecution and bodily harm. It is a sad truth that there are governments which tend to regard life cheaply, often exhibiting little consideration for the welfare of its citizens. Any society that permits serious inequalities of wealth and opportunity is prone to allow the wrong leaders to come to power, and thus cannot avoid fostering widespread resentment and frustration. It is surely no accident that the crime rate of a nation is related to environmental conditions. By far the best way to alleviate crime is to improve the social and economic system.

In one respect, it may be said that many societies deliberately condone conditions hazardous to the lives of its citizens. A rather glaring example of this would be the incredibly lax gun control laws of some countries. (For a supposedly advanced society the United States must rank as one of the least intelligent in this regard.) Why any nation would allow free access to hand guns is a great mystery, especially in view of the number of "borderline" mental cases that seem to be walking around. Why put temptation into the hands of such people?

Although most nations have recognized the great danger of drug abuse and are attempting to cope with the problem, for the sake of tax revenue they have neglected to target other insidious addictions that are detrimental to the health of all. In monetary terms alone, the carnage wrought by alcohol and tobacco greatly exceeds what revenue is collected by way of a license to play the fool. Ranging from personal injury or death at the hands of a drunken driver to a multitude of other atrocities, it is a fact that the alcoholic would be incapable of inflicting his mark upon society if that same society had never given its sanction! Similarly, if forbidden, an inconsiderate tobacco addict would not be allowed to endanger the health of innocent bystanders who are subjected to obnoxious and harmful smoke. Just as surely as though one had taken a knife and stabbed his neighbor, so a thoughtless smoker may achieve the same end result, albeit slower.

Social Reform

There are three major areas of social reform which deserve a measure of discussion— namely, justice, education and health.

Justice:

While it may be clear that an individual is obliged to follow the dictates of conscience, *it is also evident that a line must be drawn should the actions of one threaten adversity to another!* This, of course, raises the question of what to do with so-called criminals, as sin and crime are not necessarily one and the same.

As far as the individual is concerned, sin may best be defined as a *failure to obey conscience.* And yet, as we are no doubt aware, our conscience can sometimes deceive us. It may tell us that a certain action is sinful, when actually there is nothing wrong; and right when it is really wrong. It is, in fact, a variable which is ever prone to change as one evolves. *A mistaken conscience calls for correction through the application of better light, but never by violation!*

The problem of dispensing true justice is therefore seen to be one which mankind will never be able to solve completely. For one thing, we are incapable of properly evaluating the enormous effect that *environment* has upon the molding of our characters. Indeed, were the truth about this factor really known, it would be interesting to see how many prominent citizens might react to the same circumstances as were faced by others whom we would judge so unfavorably. It is conceivable that our penal institutions would be home to a surprising number of unexpected guests!

But an evaluation of sin still leaves unsolved problems. Upon what criterion should punishment be based? Exactly what is the purpose of justice? Is it primarily to exact vengeance? If so, then it is difficult to see how we could permit such a procedure and still consider ourselves to be spiritually mature. On the other hand, if the idea is to bring one to repentance and to impart a measure of spiritual enlightenment, then why is it that many souls are kept in prison (or even executed) when they have clearly come to repentance and would be useful members of society?

While there is every justification for removing individuals from society if they are a menace to the well-being of others, *it is equally important that such persons be forbidden to return to society unless there is reasonable assurance that they have overcome their weaknesses!* To allow freedom to an unrepentant criminal is just as foolish as refusal to lock up the offender in the first place. It will, therefore, be seen that the archaic practice of imposing specific "time" sentences should be discontinued, as it is wrong from both an ethical and common sense point of view. [1]

As if to add insult to injury, we willingly tolerate the spectacle of prisoners being confined for long periods of time and at very considerable public expense. Contributing nothing useful to society for many wasted years, they are all too often released as hardened individuals with a resentful attitude toward the world and little prospect for successful assimilation and gainful employment. (It is most essential that the State provide job opportunities for all released criminals, since it is highly unlikely that private enterprise will do so— thus forcing many to return to crime in order to live!) A more sensible policy would be to have them pay their way— and their victims, if at all possible— by doing useful work in State-owned industries and farms. In exchange for this labor they could learn productive skills and, in conjunction with appropriate psychiatric and spiritual counseling, eventually become an asset to society after literally earning their freedom.

As matters stand, it is really incredible that society would allow itself to be victimized twice by the same criminal. First, we have crime victims with little hope of receiving any form of compensation. Second, the general populace is saddled with the burden of paying enormous sums to maintain criminals behind bars. There is just no excuse to condone such a wasteful policy when the vast majority of inmates could be made to serve society in a useful manner. (Should there be a refusal to cooperate without legitimate cause then, quite simply, the prisoner would not eat.) Invariably, by reason of their lack of vision, it is evident that many of our highly paid legislators deserve to be replaced on grounds of plain incompetence. (In those cases where the death penalty has been prescribed, it would often make sense to work such person for life, with a "poison pill" option readily available should it be so desired.)

Education:

The vital field of education is one which invites a measure of criticism. Although the importance of a sound education is paramount in the minds of students, teachers and taxpayers alike, very few attempts seem to have been made to really assess the cost/effectiveness of our various school systems. Objections which may be raised involve both classroom curriculum and classroom attendance, along with the issue of affordability.

With regard to curriculum, it is soon apparent to many students that much time and energy is wasted on subjects of little or no practical value. (How often does one use algebra or have need for a second language, for instance?) On the other hand, many subjects of real value are mysteriously avoided like the plague. (By way of example, it is seen to be somewhat illogical for our courts to adopt a policy of "ignorance of the law is no excuse," while our school systems fail to teach any aspect of law except to university law students!)

Also ignored is the fact that technological advances have made classroom attendance very nearly obsolete. In the computer age and the Internet, why is it necessary for many students to attend a classroom when the same information could quite easily be imparted to one at home, thereby saving the taxpayer significant savings by eliminating the need for many schools, teachers and buses? In turn, such monetary savings could be used to provide more scholarships for needy students.

Yet another scholastic reform would be to allow students to progress in their studies at a rate commensurate with ability. (It is a well-known fact that many are capable of passing to higher grade levels at a much faster rate than others, and who frequently become bored and disenchanted with a school system that is quite literally holding them back.)

In the face of reluctance on the part of an archaic educational system to enter the 21st century, does one detect a fear of losing lucrative jobs— including many unnecessary over-qualified (and overpaid) educators?

Health:

Relatively few individuals are aware of a dark side to traditional medicine— a side that is actually detrimental to society in terms of both health and money. To understand the situation as it presently exists, it is important to realize that most practicing physicians today know little or nothing about nutrition! (It is simply not taught in most medical schools!) While readily conceding that some doctors have indeed performed near miracles in certain fields of health care, the sad truth is that they generally know next to nothing about a host of beneficial nutritional supplements (vitamins, minerals, etc.) that has been proven essential to good health. All the typical physician knows is focused chiefly upon prescription drugs (many of which have harmful side effects) that the giant pharmaceutical companies have persuaded them to give to their patients. The end result is that medical doctors are prescribing a multitude of potentially dangerous drugs— at highly inflated prices— when there are superior, far cheaper and much safer alternatives. (It is an established fact that most illnesses— certainly in the case of senior citizens— may be traced to nutritional deficiencies.)

Although knowledge of this nefarious state of affairs is in the process of being publicized by a handful of highly qualified medical researchers and physicians, this news has yet to filter down to the majority of misinformed doctors and their trusting patients. Hopefully, it will not be long before an aroused public demands an impartial investigation into the actions of a group of greedy pharmaceutical giants who have quite literally duped both health care agencies and physicians alike into believing in their high-priced synthetic drugs. (Since they cannot patent a natural substance, they are compelled to produce a synthetic drug and rely upon a massive publicity campaign to market their inferior products— at a huge profit, of course!)

As incredible as it may seem, these same pharmaceutical giants have actually banded together and are attempting to make almost all beneficial natural supplements a prescription item— thereby depriving the public of their right to obtain a superior and essential product at far lower cost! Such is the public mischief wrought by a conglomeration of greedy pharmaceutical companies operating under the guise of benefitting society! (By way of example, inexpensive colloidal silver— pure silver held in fine suspension— may be cited as a substance that is superior to the vast majority of antibiotics, and has the added benefit of being effective against a wide range of deadly viruses. Moreover, it has no adverse side effects and neither bacteria nor viruses have been known to mutate against it to produce new drug-resistant strains.)

Regrettably, there is a rarely discussed health concern, of global significance, which now looms on the horizon— one that threatens mankind from *within*. Thanks to the efforts of medical science, the lives of many individuals, plagued with a variety of serious genetic diseases (i.e., diabetes, hemophilia, cystic fibrosis, sickle cell anemia, etc.) Have been greatly extended. While this is certainly to be desired, it has nevertheless interfered with nature's evolutionary process by allowing many carriers of defective genes to live past the age of puberty— thereby passing on these defects to succeeding generations. The end result of this intervention is *cumulative* in effect, and it will not be much longer than mere centuries before a large proportion of humanity will carry a significant genetic defect! What is presently needed is not so much a marriage license as a birth license, before the situation gets totally out of hand. In conjunction with a worldwide policy of birth control (needed to contain an exploding population), this weeding out of genetic defects is a problem which cannot be avoided indefinitely and must be addressed effectively— preferably upon the level of a reformed United Nations— in the very near future, and before quantity is allowed to eclipse quality.

Economic Reform

Contrary to popular opinion, the natural resources of our planet rightfully belong to everyone— not to those whom same arbitrary national boundary would seem to bestow a gift. If ever we are to rise above the mentality of ignorant tribesmen, it is essential to realize that no nation has a legitimate and exclusive claim to that which has been supplied by nature. The selfishness of a few to the detriment of many is an attitude that can no longer be condoned by a civilized humanity. It is just as much an outrage for a group of oil-rich sheiks to resort to acts of extortion, as it is for other nations to refuse food to a starving populace. Similarly, the timber, fisheries and mineral resources of the world are a heritage given to all. The ability of nations to share their blessings is a must if there is to be a peaceful resolution of global tensions.

The issue of economic reform invariably raises the question of public ownership. To what extent is it desirable to replace private enterprise with a system of democratic State ownership? In spite of the numerous injustices which capitalism has spawned, it must not be thought that the idea of free enterprise is without merit. On the contrary, inequalities arose chiefly because there has been no effective limitation upon the wealth and power accumulated by a small minority of citizens; and as long as a more just distribution of wealth can be assured there are many features that are beneficial in our current stage of social evolution. The right of one to own a home, farm or business of reasonable size should be recognized by the State.

Nevertheless, it is also clear that a considerable amount of public ownership is essential if present objections are to be overcome and resources utilized for the benefit of all. In particular, there must be a limitation placed upon personal wealth, as it is this great inequality that leads to so many undesirable consequences. But what specific business enterprises should be owned by the public, and what actual restrictions should be placed upon private wealth and business corporations?

Limitation of wealth may be achieved by placing an initial ceiling upon the total assets that may be held by an individual. The exact figure of this proposed limitation is something which must obviously be based upon circumstances and the wealth of a nation. However, a tentative value of perhaps thirty times the yearly income of the average worker is not entirely out of line with the need to encourage private initiative on the one hand, and the necessity to achieve a more acceptable distribution of wealth on the other. Future injustice could be negated through imposition of higher succession duties, and by an income tax so designed as to confiscate entirely all income above a certain sum— possibly an amount equal to no more than ten times the average annual wage. (The structure of such tax laws would, of course, be founded on the *progressive tax principle*: essentially, the greater the income the higher the percentage of tax— with a basic exemption sufficient to exclude those of low income.)

Public control over a portion of a nation's economy will serve, in many ways, to regulate that portion under private ownership, since any private business will be competing with the State for available labor. Thus wages and working conditions must be competitive with businesses under public domain. Except for whatever limitation may be imposed by private inability to finance a really large business, a free economy will prevail and inspire the greatest possible efficiency. If public control of a particular industry should prove more efficient and more desirable than private control, the former will eventually replace the latter on grounds of merit. Conversely, if private initiative can be shown to be equal or superior in a specific business, then there is no reason to change to State ownership.

By far the most important business demanding public ownership is that of *finance— a medium which totally regulates the very economy of a nation!* It is incredible that anything so vital should be entrusted to private ownership. Future generations will surely be amazed by policies which we presently condone. Not only does private control of finance often result in a poor channeling of funds to worthy projects, but the established practice of paying interest for the use of money constitutes an unnecessary hardship for many, and serves to lower the standard of living for all but the wealthy. The idea of charging interest for borrowed money has little justification, either ethically or from the point of view of economic stability. Morally speaking, any money obtained by way of interest cannot be considered earned, as no genuine contribution to society will have been made. On economic grounds, there is certainly no conflict with the requirement that the money supply must always balance available consumer's goods, since all loans must be repaid. The concept of demanding interest in return for the use of money constitutes one of the main principles upon which the capitalist system is based. In effect, it is actually little more than a device to place an unnecessary burden upon the shoulders of the public. The only ones to gain are those who possess wealth!

Admitting, therefore, that the principle of usury is unjust and undesirable, why is it condoned by society? Inexplicably, there would appear to exist a widespread belief that there is actually a necessity to borrow money from private sources. To illustrate this ridiculous obsession we have only to observe the various levels of our government— *for even they have somehow fallen into the clutches of private financiers!* Instead of simply printing and loaning money (which must be repaid) free of interest to itself for worthwhile purposes, we witness the absurd spectacle of our government first permitting such funds to pass through the hands of private financial institutions, who very obligingly seize upon this "windfall" opportunity to ask interest. Thus there exists a situation whereby sanction from government is necessary to print money, but in order for the various levels of government to obtain these funds it must consent to pay "middlemen" large sums of money in the form of unnecessary interest to give it back to itself! How crazy can we get?

The abolition or restriction of usury would result in enormous benefits to municipalities and private citizens alike. As a direct consequence of State control of finance, money for constructive municipal and state purposes would be available for only minimal administrative fees, thereby reducing costs and taxes. Of immense benefit to the average individual would be the benefit of being able to secure virtually free loans for the erection or purchase of homes, farms and small businesses.

Yet another form of injustice, inherent in the monetary policies of capitalism, is the practice of deficit financing. It is a fact of sound economics that stability requires a balanced national budget. Quite simply, there is no excuse to let the money supply get out of synchronization with available consumers' goods. In times of war or rearmament, when surplus money is injected into the system, the usual procedure is to avoid recovering the full amount of justified tax— thereby producing inflation, which is alleviated momentarily through the supposedly patriotic gesture of issuing government savings bonds. This is surely a most despicable ploy that should be outlawed by constitution! *It is nothing less than a crooked device to allow the wealthy to escape their due responsibilities*, since it is obvious that the poor cannot be asked to pay substantially more than they already have without risking strong reaction. By this blatant subterfuge, the rich are permitted to retain excess profits and a "debt" is incurred for which there is absolutely no justification. (From an economic standpoint, it is clearly impossible for a nation to have a legitimate debt for something *already produced with its own resources!*)

The mentality of Earth politics would never cease to amaze a visitor from another planet. While many capitalist nations finally got around to initiating a variety of old age pension and unemployment insurance schemes, in order to compensate for loss of income, for some inexplicable reason they have all failed to see the merit of *adjusting retirement age* as a means of abolishing unemployment altogether! By merely lowering the age at which one could retire on pension, in times of unemployment, it would free sufficient jobs as to totally eliminate the problem. Why endure the common spectacle of disgruntled unemployed youth, torn between idleness and temptation to turn to crime, while many older workers— desirous of retiring , if only they could afford to do so— continue to labor and pay taxes? In effect, older and less healthy individuals are compelled to support those more capable and more desirous of membership in the labor force. In no way can this stupid oversight be excused on grounds of cost, since the additional amount required for pensions need never exceed what is currently being spent on such props as unemployment benefits, welfare payments, and a host of essentially useless retraining and "make work" programs— of which enticing (or forcing) youth into the military may be cited as a modern approach to the problem.

A planned economy, devoid of any unfair distribution of wealth, wasteful duplication, unemployment and (above all) senseless military expenditures, would at last permit genuine peace and a great increase in world living standards.

Global Unification

Almost since its inception, it has been apparent that the United Nations, as currently formulated, is far too weak and ineffectual to guarantee world peace. In many respects a carbon copy of its failed predecessor, the League of Nations, this body has proven to be no more successful in resolving international disputes and in achieving world disarmament. A multitude of wars and atrocities have been allowed to ravage the globe without effective interference, in open defiance of its very presence! The momentous question now confronting us is whether there is any way in which it may be revitalized and given the power to deal with aggression and injustice.

A new United Nations:

By far the most pressing issue facing the world today involves the abolition of nuclear weaponry and germ/chemical warfare. Taken either separately or together, these twin evils threaten the annihilation of Earth society. It is a highly ominous state of affairs to realize that a multitude of minor nations possess (or will soon possess) one or more of these fearful capabilities! What makes the situation even more alarming is that history has repeatedly demonstrated man's willingness to resort to acts of sheer terrorism, sometimes with little provocation or forethought as to consequences. It requires no great taxing of the imagination to envisage the wholesale carnage once some mentally unbalanced leader (and there always seems to be a number of such characters in power) begins pushing buttons. Following an inevitable retaliation, the stage would be set for uncontrolled escalation. The time to establish a *World Constitution*, which would positively— and effectively— outlaw weapons of this nature, is now, before it becomes too late! To delay is to risk global suicide.

This procrastination in adopting an effective disarmament policy is most distressing. Can it not be seen that further delay will soon result in a rapidly growing list of small nations (many with dictatorial leaders) being able to deliver a staggering amount of mass destruction? To compound the problem, once a minor nation acquires the Bomb (which can be delivered very effectively and cheaply by means of a "suitcase") an unscrupulous dictator will have little incentive to give it up and rely upon mere conventional arms— a field in which it cannot compete!

Based upon proportional representation among those nations which embrace the democratic principle, with no nation possessing the power of veto, the foundation would at last be laid for a *World Government* to devise a prompt and equitable solution to many of Earth's problems. The time has come for mankind to put aside silly grievances and prejudices, which lie far beneath the demands and responsibilities of the 21st century. What is needed is a spirit of cooperation and mutual trust, in which individuals of all creeds and nationalities are willing to grow up and *think unselfishly in terms of the entire human race!* Unification— not separation— must guide our every thought and action.

Comprised of carefully screened and motivated volunteers from many nations— with no ties to religious dogma and vain nationalism— and pledged to fight aggression and injustice wherever it is seen to exist, a *World Army* could be formed that would do much to assure global peace and prosperity. With the overwhelming superiority of such a force, a revitalized United Nations would, for the first time in its history, be in a position to do something about the many injustices posed by our planet's assorted petty dictators. A top priority of this International Police Force would be to initiate a campaign to overthrow all governments which fail to embrace the principle of democracy. Further, it would strive to remove from power those governments which seek to impose any form of religious dogma upon its citizens. In essence, there must be complete freedom to chose one's religion— without having a set of beliefs thrust upon them.

The Issue of Terrorism:

Regrettably, our inept United Nations, which has tolerated a host of atrocities— with only feeble attempts to punish the guilty— has led to unprecedented acts of terrorism. When unscrupulous madmen note that there is little more than verbal reaction to their carnage, they are encouraged to continue their ways. (The fact that nefarious dictator Saddam Hussein was allowed to stay in power after the First Gulf War is almost beyond belief!) Clearly, the wrong message is being sent to all would-be-dictators— namely, that no matter what the crime, a country's leader will not be punished. How crazy can we get?

But the actual perpetrators of acts of violence are not always the ones who are the most guilty. Many of the Middle East "suicide bombers" have been quite literally "brainwashed" into committing their acts of terrorism. In a sense, the real culprits are those cowardly religious fanatics who have conned others to do their dirty work. Reluctant to give up their own lives, they would incite violence by cajoling weak-minded followers to do their bidding. Such unprincipled religious fanatics must be eliminated— by whatever means it may take— as they are the real enemy to be eradicated. Indeed, one who would advocate violence is just as guilty as one who commits the actual crime. Accordingly, such an instigator deserves the same punishment. (It would be interesting to see how long it would take to stop acts of wanton terrorism if a significant ratio of convicted and unrepentant militants were to be executed in response to the murder of innocent victims, soldier or civilian alike!)

A Universal Language:

A rather pertinent question arises as to why our present United Nations, supposedly devoted to the task of uniting nations, has not even pressed for a universal language. If ever there was a factor designed to prevent humanity from achieving harmony, it is surely the lack of a common speech. But why continue this foolish babbling, in multiple tongues, when there is absolutely no reason to tolerate such a ridiculous scenario? It stands as little credit to our witless politicians that they have not seen fit to insist upon a *world language!* Derived from the best features of many cultures, a scientifically based language could easily be devised and taught in the schools of every nation. Within a generation or two it could become the official tongue of the planet Earth, supplanting all others and providing a new basis for world understanding. It would also signify that, at long last, we have reached the stage where we might be justified in thinking of ourselves as prospective members of a Galactic Society—rather than a disorganized rabble which has yet to evolve beyond a primitive state of humanity.

Hopefully, mankind will soon awaken to the challenges which threaten the evolution of our species, as time is not unlimited before our planet's inevitable rendezvous with Cosmic Destiny.

Note:

[1] A number of glaring faults may be cited with regard to our present court system. First, new evidence must always be allowed to remedy a wrongful conviction, with compensation given for any unjust incarceration. Second, bias in favor of the rich needs to be

addressed. (The very practice of posting bail favors the wealthy.) Why, in fact, should one be required to hire a lawyer (often for an exorbitant fee) in order to obtain justice? Clearly, the rich can afford the best counsel. What is needed is a court system in which all are given an equal opportunity to state their case, and without any necessity to literally buy justice!

Addendum to *"Economic Reform"*:

The following text in "Letters to the Editor" was published March 6, 2009 in the Hernando Today issue of the Tampa Tribune"

Try "Trickle-Up"

It is surprising how little our politicians know about economics. A sound economy is only possible if the people have sufficient money to purchase available consumer goods. Any "trickle-down" policy will simply fail, since money injected at the top tends to stay at the top and does not address the real problem— namely, lack of purchasing power at the bottom.

Recessions and depressions are due to insufficient funds in the hands of the masses. Consumer goods (homes, cars, etc.) are available, but the public cannot afford them. Bluntly put, any stimulus package must be given— as an outright gift— to the workers. If sufficient funds are given to the people, they will quickly pay off debts and purchase the goods— thereby enabling factories and builders to rehire employees.

By simple arithmetic, a trillion dollar stimulus program is equivalent to more than $10,000 dollars for every household in America! Would it not make far more sense to give this money to the people themselves, rather than to the rich business leaders who got us into such a financial mess?

Harold W.G. Allen